Endocrine Development

Vol. 23

Series Editor

P.-E. Mullis Bern

Developmental Biology of GH Secretion, Growth and Treatment

Volume Editor

Primus-E. Mullis Bern

22 figures, 1 in color, and 10 tables, 2012

Basel · Freiburg · Paris · London · New York · New Delhi · Bangkok · Beijing · Tokyo · Kuala Lumpur · Singapore · Sydney

Endocrine Development

Founded 1999 by Martin O. Savage, London

Primus-E. Mullis
Paediatric Endocrinology
Diabetology and Metabolism
University Children's Hospital
Inselspital
Bern, Switzerland

Library of Congress Cataloging-in-Publication Data

ESPE Advanced Seminar in Developmental Endocrinology (6th : 2012 : Bern, Switzerland)
  Developmental biology of GH secretion, growth, and treatment / 6th ESPE Advanced Seminar in Developmental Endocrinology, Bern, May 10-11, 2012 ; volume editor, Primus-E. Mullis.
      p. ; cm. -- (Endocrine development, ISSN 1421-7082 ; vol. 23)
  Includes bibliographical references and indexes.
  ISBN 978-3-318-02244-5 (hard cover : alk. paper) -- ISBN 978-3-318-02245-2 (electronic version)
  I. Mullis, P.-E. II. Title. III. Series: Endocrine development ; v. 23.
1421-7082.
  [DNLM: 1. Growth Hormone--physiology--Congresses. 2. Growth Hormone--secretion--Congresses. 3. Growth Hormone--deficiency--Congresses. 4. Human Development--physiology--Congresses. W1 EN3635 v.23 2012 / WK 515]

  612.4'92--dc23

                                        2012034041

Bibliographic Indices. This publication is listed in bibliographic services, including Current Contents®.

© Copyright 2012 by S. Karger AG, P.O. Box, CH–4009 Basel (Switzerland)
www.karger.com
Printed in Germany on acid-free and non-aging paper (ISO 97069) by Kraft Druck GmbH, Ettlingen
ISSN 1421–7082
e-ISSN 1662–2979
ISBN 978–3–318–02244–5
e-ISBN 978–3–318–02245–2

Contents

Preface

In the last few years, rapid progress has taken place in our understanding of the 'developmental biology of growth hormone (GH) secretion' and its pivotal role in growth. To keep the reader up dated on these most important developmental aspects and influences leading to changes in terms of clinical views, we focus first on 'pituitary gland development and imaging'. In these two chapters Prof. Mehul Dattani highlights the various factors most important for pituitary gland development causing pituitary gland-derived hormonal deficiencies and its impact on clinical phenotypes, whereas Prof. Mohamad Maghnie describes the power of imaging approaches on diagnostic procedures and following the suggested genotype which might help in finding the appropriate diagnosis easier, and, therefore, to save time and, very importantly, money. Coming to the 'genetics of the GH axis', Profs. Jan-Marteen Wit and Roland Pfäffle concentrate on the spectrum of isolated insulin-like factor-1 (IGF-1) deficiency (IGFD) as well as on the molecular defects downstream of IGF-1. Dr. Martin Bidlingmaier, a leading export on GH detection, not only as far as doping is concerned, emphasizes the importance of the new detection methods of GH and growth factors. What would such a meeting/seminar be without the two most important topics, 'epigenetics' and 'bioinformatics', covered by Prof. Irène Netchine and PD Dr. Amit Pandey. By going more and more in-depth, Prof. Mike Waters' contribution stresses the new aspects of growth hormone and cell growth. At the very end, Drs. Petkovic and Miletta round the topic out by adding two articles on the cutting-edge of timely research.

Thanks to the endeavors of the European Society of Paediatric Endocrinology, this 6th Advanced Seminar in Developmental Endocrinology dedicated to physicians and basic scientists in the field of endocrinology was held in Bern, Switzerland, in May 2012. This seminar gathered 30 participants from all over the world to present their data and patients, and to debate with distinguished international investigators and scientists in the field.

Proceedings of meetings and seminars are very often out of date, old and, therefore, difficult to appreciate. Therefore, to overcome this conflict and criticism, the aim was to have it published without any great delay and to keep it on the front-line of actuality. I strongly believe that at this point we were very successful and that this splendid collection of investigation data and reviews will without doubt serve as a reference for all pediatricians and scientists interested in growth and development.

Primus-E. Mullis, Bern

Mullis P-E (ed): Developmental Biology of GH Secretion, Growth and Treatment.
Endocr Dev. Basel, Karger, 2012, vol 23, pp 1–15 (DOI: 10.1159/000341733)

Pituitary Gland Development: An Update

Rodrigo E. Bancalari · Louise C. Gregory · Mark J. McCabe ·
Mehul T. Dattani*

Developmental Endocrinology Research Group, Clinical and Molecular Genetics Unit, University College
London-Institute of Child Health, London, UK

Abstract

The embryonic development of the pituitary gland involves a complex and highly spatio-temporally
regulated network of integrating signalling molecules and transcription factors. Genetic mutations
in any of these factors can lead to congenital hypopituitarism in association with a wide spectrum of
craniofacial/midline defects ranging from incompatibility with life to holoprosencephaly (HPE) and
cleft palate and septo-optic dysplasia (SOD). Increasing evidence supports a genotypic overlap with
hypogonadotrophic hypogonadal disorders such as Kallmann syndrome, which is consistent with
the known overlap in phenotypes between these disorders. This chapter reviews the cascade of
events leading up to the successful development of the pituitary gland and to highlight key areas
where genetic variations can occur thus leading to congenital hypopituitarism and associated
defects.   Copyright © 2012 S. Karger AG, Basel

The pituitary gland is a midline structure located within the sella turcica recess of
the sphenoid bone at the base of the brain. The success of its development during
embryogenesis is dependent upon complex interactions of spatio-temporally regulated signalling molecules and transcription factors. The final organ is composed of
three lobes which are derived from dual-embryonic ectodermal origins; the endocrine hormone-producing anterior and intermediate lobes originate from the oral
ectoderm and the posterior lobe develops from the overlying neural ectoderm [1].
Maintained apposition and interactions between these ectodermal layers is critical for
normal pituitary organogenesis.

Defects in this developmental process can result in the loss or reduction of
hormone-secreting cells thus causing congenital hypopituitarism, the pathology of
which can vary from deficits in multiple pituitary hormones (combined/multiple
pituitary hormone deficiency, CPHD/MPHD) or a single hormone in isolation, the
most common deficit being in growth hormone [2].

The midline location of the pituitary gland and its close developmental associations with forebrain and eye development often results in the association of congenital hypopituitarism with highly heterogeneous midline defects ranging from incompatibility with life, to holoprosencephaly (HPE) and cleft palate and septo-optic dysplasia (SOD), which will be described later [3].

The scope of this chapter is to provide a background on the development of the pituitary gland and the signalling molecules and transcription factors which drive this process. The genetic abnormalities which lead to associated defects such as HPE and SOD will also be described plus an update on the apparent overlap between these groups of disorders and Kallmann syndrome, defined as the combination of hypogonadotrophic hypogonadism (HH) and anosmia.

Pituitary Gland

Morphology

After the initial dorsal invagination of the primordium of the anterior pituitary lobe from the oral ectoderm (see below) to form Rathke's pouch (RP), a series of tightly regulated cellular proliferation and differentiation events eventually give rise to five highly differentiated cell types secreting a total of six hormones: (1) adrenocorticotrophic hormone (ACTH) from the corticotrophs, (2) thyrotrophin or thyroid stimulating hormone (TSH) from the thyrotrophs, (3) growth hormone (GH) from the somatotrophs, (4) prolactin from the lactotrophs, and (5) follicle-stimulating hormone (FSH) and luteinising hormone (LH) from the gonadotrophs [1]. The intermediate lobe, derived from the same invagination event, contains melanotrophs which secrete pro-opiomelanocortin (POMC), a major precursor protein to endorphins and melanocyte-stimulating hormone (MSH) [2]. Secreted hormones then interact with distinct targets throughout the body. The ventral diencephalon (VD) (primordial hypothalamus) component of the neural ectoderm evaginates ventrally to form the infundibulum (see below), which is destined to form the posterior pituitary and pituitary stalk, the latter structure acting as a physical connection between the pituitary gland and brain and containing the hypophyseal (hypothalamo-pituitary) portal system as well as neural connections between the hypothalamus and posterior pituitary [4].

The posterior lobe is comprised of axonal projections of neurons which traverse the pituitary stalk and median eminence at the base of the hypothalamus. The neurons originate from hypothalamic magnocellular bodies termed the supraoptic, suprachiasmatic and paraventricular nuclei. The former two nuclei release arginine vasopressin whilst the paraventricular nuclei release oxytocin [4]. The median eminence contains a capillary bed into which parvocellular neurons, located in a variety of hypothalamic nuclei, secrete hypophysiotropic hormones which then reach the anterior and intermediate pituitary lobes via the hypophyseal portal system and stimulate the release of anterior/intermediate lobe hormones. Thus, it is the hypothalamus

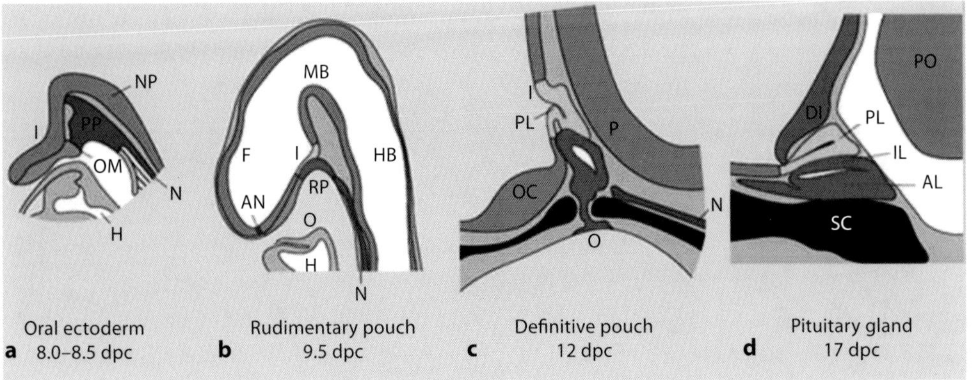

Fig. 1. Development of the murine pituitary. Thickening of the stomodeum (oral ectoderm) at E8.5 marks the onset of pituitary organogenesis (**a**), which is followed 1 day later by the invagination of a rudimentary pouch toward the overlying ventral diencephalon (**b**). The definitive Rathke's pouch is formed as its connection to the oral ectoderm is severed and the infundibulum evaginates from the ventral diencephalon to form the posterior pituitary (**c**). Progenitors of the hormone-secreting cell types then proliferate and terminally differentiate to produce the mature pituitary gland consisting of the anterior lobe, intermediate lobe and posterior lobe (**d**). E = Embryonic day; I = infundibulum; NP = neural plate; N = notochord; PP = pituitary placode; OM = oral membrane; H = heart; F = forebrain; MB = midbrain; HB = hindbrain; RP = Rathke's pouch; AN = anterior neural pore; O = oral cavity; PL = posterior lobe; OC = optic chiasm; P = pontine flexure; PO = pons; IL = intermediate lobe; AL = anterior lobe; DI = diencephalon; SC = sphenoid cartilage. Taken from Sheng et al. [12], with permission.

which acts as the central regulator of growth, reproduction and homeostasis via the pituitary gland [4].

Timeline of Pituitary Organogenesis
Neurulation occurs at 3 weeks of gestation in humans and is the first known event of nervous system development [3]. This involves the thickening of the notochord to give rise to the neural plate (primordial central nervous system) from which the brain and spinal cord are derived. The anterior neural plate is the predecessor of the forebrain, optic nerves, hypothalamus and anterior and posterior pituitary lobes. The development of these latter structures has been well characterised in murine models [4] and is highly conserved across vertebrate species including humans. As such, the following paragraphs will outline murine pituitary organogenesis.

As embryogenesis proceeds, the growth and expansion of the brain causes the head of the embryo to bend anteriorly, displacing the heart ventrally and causing a depression between the heart and brain which is termed the stomodeum, or oral ectoderm. The thickening of the upper edge of the stomodeum at embryonic (E) day 8.5 marks the onset of pituitary organogenesis (fig. 1). At E9.5, the thickened section of stomodeum invaginates to form the rudimentary RP. Twenty-four hours later, the infundibulum evaginates from the VD located within the neural ectoderm (which had thickened at E7.5) to come into contact with RP. This essential contact is maintained throughout pituitary organogenesis. At the same time, the connection between RP and the oral ectoderm is severed with the generation of a fully developed definitive pouch. From E12.5–E17.5 the progenitors of the five hormone-secreting cell types proliferate and terminally differentiate ventrally, while the dorsal portion of the pouch becomes the intermediate lobe, the size of which varies greatly between species and is largely absent in humans.

The first hormone-secreting cell types appear at the most ventral portion of RP at E11.5 after expressing the α-glycoprotein subunit (*αGSU*) [4]. These same cells express transcription factor islet-1 (*Isl1*) which corresponds to prospective thyrotrophs. These cells terminally differentiate at E12.5 after expressing thyroid-stimulating hormone subunit β (*Tshβ*), although these cells are short lived and disappear at birth. Dorsal to these intermediary thyrotrophs are corticotrophs which begin differentiating by E12.5. They express *Pomc*, as do the melanotropes from E14.5 in the intermediate lobe. Around the same time definitive thyrotrophs are also detected in the anterior pituitary lobe [4]. At E15.5 somatotrophs and lactotrophs begin differentiating with the detection of growth hormone and prolactin, respectively. The former cell types then proliferate through the central and lateral aspects of the anterior lobe [4]. Lactotroph localisation is more restricted to the medial zone, adjacent to the intermediate lobe. The gonadotrophs are the last anterior pituitary cell types to be detected with Fsh sub-unit β (*Fshβ*) detected at E16.5, followed by Lh subunit β (*Lhb*) one day later [4].

Molecular Regulation of Pituitary Development
This section details the molecular regulation of murine pituitary development and how this fits with the timeline above, and is illustrated in figure 2.

Factors Involved in the Early Formation of Rathke's Pouch
Bone Morphogenetic Protein 4 and the Sonic Hedgehog Pathway
Bone morphogenetic protein 4 (BMP4) is the earliest signalling molecule known to be expressed in the prospective infundibulum at E8.5, as the stomodeum thickens to form the presumptive RP. As such, Bmp4 is required for the initiation of RP forma-tion and its continued expression through to E14.5 suggests it is important for the maintenance of RP formation also. Complete deletion of *Bmp4* usually results in early embryonic lethality [4], although analyses of the few mice which survive to E10 show no signs of RP formation, or even a thickening of the stomodeum.

In humans, *BMP4* is located on chromosome 14q22–q23 near the *OTX2* gene [5, 6]. During embryogenesis *BMP4* is expressed in the optic vesicle and optic cup, which is consistent with its known expression across species prior to lens formation. In the brain *BMP4* is expressed in the diencephalic floor, consistent with its role in pitu-itary development and also the medial ganglionic eminence. *BMP4* is also expressed in developing limbs [5, 6]. Various intragenic mutations including missense (heterozy-gous, compound heterozygous and homozygous), nonsense, chromosomal deletions and frameshift variations have been detected across multiple patients with various

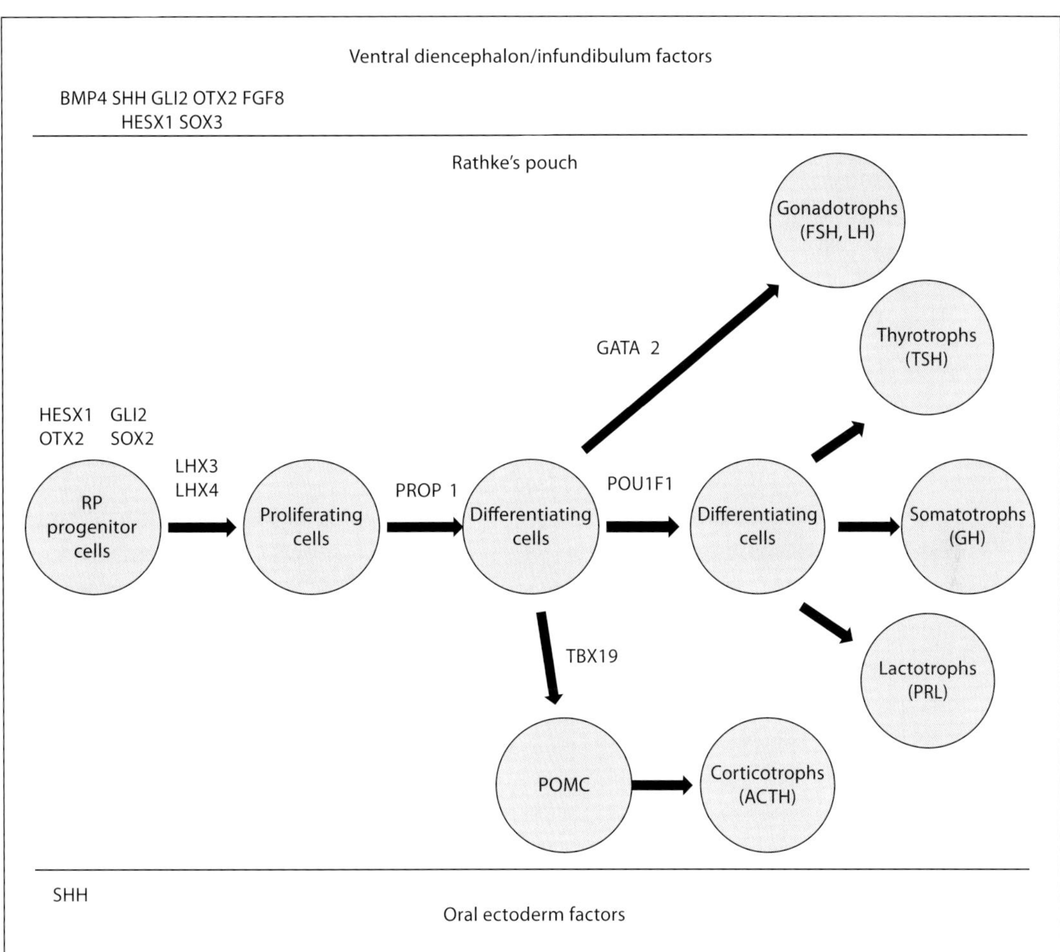

Fig. 2. Schematic representation of Rathke's pouch (RP) development from the earliest progenitor cells through to the five hormone-secreting cell types. The transcription factors involved in regulating this process are expressed from the (1) ventral diencephalon or infundibulum (top), (2) oral ectoderm (bottom) or (3) from within the pouch itself. Note this Figure is a highly simplified version of a much more complex process with the factors listed herein being those described within the text and not intended to be an exhaustive list. ACTH = Adenocorticotrophic hormone; FSH = follicle-stimulating hormone; GH = growth hormone; LH = luteinising hormone; POMC = pro-opiomelanocortin; PRL = prolactin; TSH = thyroid stimulating hormone.

phenotypes including anophthalmia/microphthalmia, congenital glaucoma and sclerocornea (eyes), decreased brain white matter, lateral ventricular dilatation, hypoplastic corpus callosum, hydrocephaly, cleft lip and palate, developmental and growth delay (brain and craniofacial/midline structures) and postaxial polydactyly, syndactyly and brachydactyly (limbs) [5]. Interestingly, the literature is deficient on the state of the pituitary gland in most patients except for one case involving a chromosomal deletion of 14q22–q23 encompassing *BMP4, OTX2* and *SIX6* [6]. All three proteins are expressed in areas critical for pituitary development and as such it is not clear which of these proteins were causative of the pituitary phenotype. Thus further clarification of the pituitary phenotype in cases of isolated *BMP4* mutations/deletions is required.

BMP4, along with FGF8 (see below), appears to be involved in the spatially-restricted distribution of Sonic Hedgehog (SHH) expression within the oral ectoderm. In mice, genetic expression of *Shh* is extensive throughout the VD until E14.5 and oral ectoderm except for RP until E12.5. Shh is a member of the hedgehog family of morphogens which elicit different cellular responses and cell fates. It binds to the transmembrane protein Patched which subsequently dissociates from its co-receptor Smoothened thereby activating the GLI family of transcription factors, which are involved in activating or repressing target genes. Patched and the GLI transcription factors are expressed within RP (with the latter also being present in the VD) meaning that Shh can mediate pituitary development through these pathways [4]. Due to the severity of the mouse *Shh*-null phenotype which exhibits cyclopia and loss of brain midline structures, the pituitary phenotype cannot be assessed. However, mice which are transgenic for an inhibitor of the hedgehog family do develop RP due to the normal expression of *Bmp4* and *Fgf8*, but it is severely hypoplastic. Humans with mutations in *SHH* exhibit holoprosencephaly (HPE), which is the failure of the brain to divide into its two hemispheres, often accompanied by craniofacial malformations and almost always developmental delay. Approximately 20 *SHH* mutations have been identified in HPE patients, with at least 8 mutants presenting dramatic impairment in the production of the active SHH N-terminal [7]. No *SHH* mutations have yet been discovered in patients with hypopituitarism either in isolation or in association with other defects. Similarly, human patients with mutations in *GLI2* also present with HPE in conjunction with craniofacial/midline abnormalities such as a single central incisor, single nares, optic nerve hypoplasia, partial agenesis of the corpus callosum and pituitary gland dysfunction as well as postaxial polydactyly. There have been 17 *GLI2* mutations identified in such patients with variable phenotypes (table 1), often with panhypopituitarism or isolated growth hormone deficiency. Recently, mutations in GLI2 have been associated with hypopituitarism in the absence of holoprosencephaly [8]. Loss-of-function studies show that the homozygous *Gli2* knockout mouse is perinatal lethal with severe cartilage and bone developmental abnormalities [9].

FGF8

Fibroblast growth factor 8 (Fgf8) is one member of the large fibroblast growth factor family (total of 23 members). It is expressed in the infundibulum by E9.5 and has been shown to regulate the expansion of RP through induction of *Lhx3* and *Lhx4* expression (see below) in the pouch [10]. In *Fgf8* hypomorphic mice, the anterior pituitary gland is markedly hypoplastic and the posterior pituitary may be absent [10]. Midline defects including HPE were present in the mutant mice. The arginine vasopressin (AVP) and oxytocin neurons of the hypothalamus are also reduced in number; these data were consistent with the diabetes insipidus observed in an autosomal-recessive case of HPE due to a homozygous mutation in *FGF8* [10]. Mutations in this gene were only recently

Table 1. Mutational characteristics of transcription factors mediating pituitary development

Gene	Inheritance	Types	Endocrinopathy	Phenotype
GLI2	haploinsufficiency	missense, frame-shift	CPHD (GH, TSH, LH, FSH, ACTH)	HPE, craniofacial abnormalities, polydactyly, HH, partial ACC
FGF8	AR, AD	missense, chromosome deletion	LH, FSH, DI	HH, anosmia, HPE, Moebius syndrome, SOD
LHX3	AR	missense, nonsense, frame-shift, splicing	CPHD (GH, TSH, LH, FSH, PRL, ACTH)	limited neck rotation, short cervical spine, sensorineural deafness
LHX4	AD	missense, frame-shift	CPHD (GH, TSH, ACTH), variable GnD	cerebellar abnormalities
HESX1	AR, AD	missense, frame-shift	IGHD or CPHD	SOD
SOX2	AD	missense, nonsense, frame-shift	LH, FSH, variable GHD	anophthalmia/ microphthalmia, oesophageal atresia, genital tract abnormalities, hypothalamic hamartoma, sensorineural hearing loss, diplegia
SOX3	XL	variations in polyalanine tract length, chromosome duplication	IGHD or CPHD	mental retardation, infundibular hypoplasia, EPP, midline abnormalities
OTX2	AD	missense, nonsense, microdeletion	IGHD or CPHD	anophthalmia/ microphthalmia, coloboma, developmental delay
PROP1	AR	missense, nonsense, frame-shift, splicing	CPHD (GH, TSH, LH, FSH, PRL), evolving ACTH deficiencies	may show transient AP hyperplasia
POU1F1	AR, AD	missense, nonsense, frameshift, splicing	CPHD (GH, TSH, PRL)	variable anterior pituitary hypoplasia
TBX19	AR	missense, nonsense, frameshift, splicing	ACTH	neonatal hypoglycaemia, neonatal cholestatic jaundice

ACC = Agenesis of the corpus callosum; ACTH = adenocorticotrophic hormone; CPHD = combined pituitary hormone deficit; EPP = ectopic posterior pituitary; FSH = follicle-stimulating hormone; GH = growth hormone; GHD = growth hormone deficiency; GnD = gonadotrophin deficiency; HH = hypogonadotrophic hypogonadism; HPE = holoprosencephaly; IGHD = isolated growth hormone deficiency; LH = luteinising hormone; PRL = prolactin; SOD = septo-optic dysplasia; TSH = thyroid-stimulating hormone.

described in patients with phenotypes associated with congenital hypopituitarism, including the above-mentioned patient with autosomal recessive HPE (table 1) [10].

Lim Homeodomain Transcription Factors

Lhx3 is expressed early during anterior pituitary development, initially with strong uniform expression within RP at E9.5 [11] as well as in the ventral hindbrain and spinal cord. Expression persists throughout the adult pituitary, suggesting a maintenance function in one or more of the anterior pituitary cell types [1]. *Lhx3*$^{-/-}$ mice die shortly after birth and present with defects in the differentiation of all pituitary hormone secreting cell-types despite the initial formation of RP. Mutations in *LHX3* have now been identified in a number of human pedigrees characterized by combined pituitary hormone deficiency and cervical abnormalities with or without restricted neck rotation; some patients also exhibit sensorineural hearing loss (table 1) [4]. Magnetic resonance imaging reveals a small or enlarged anterior pituitary with a structurally normal posterior pituitary. In some patients, a microadenoma may be observed.

Closely related to *Lhx3* is *Lhx4* which is also expressed throughout the invaginating RP at E9.5. Its expression, however, is transient and becomes restricted to the future anterior lobe by E12.5 and eventually downregulated at E15.5 [12]. Loss of *Lhx4* does not prevent definitive pouch formation but results in the formation of a hypoplastic pituitary, containing each of the differentiated cell types albeit at lower numbers compared to wild-type mice (in contrast to *Lhx3* mutants). Mice die shortly after birth. Patients with *LHX4* mutations have variable phenotypes including GHD and variable TSH, gonadotrophin and ACTH deficiencies with a hypoplastic anterior pituitary, with or without an ectopic posterior pituitary (table 1) [1].

Another family member, *Lhx2*, which is expressed in the VD, infundibulum and posterior pituitary, had been known to cause anophthalmia and malformation of the cerebral cortex when disrupted in mice. More recently, deletions in *Lhx2* were demonstrated to disrupt the morphology and organisation of RP despite all endocrine cell lineages being present [3]. No mutations in humans with hypopituitarism have been detected although one heterozygous variant (p.P43A), predicted to be significant, was recently detected in a patient with anophthalmia, although the paternal carrier was phenotypically normal [13].

HESX1

Homeobox embryonic stem cell 1 (*HESX1*) is a member of the *paired-like* homeobox gene family which encodes transcription factors that contain a DNA-binding homeodomain. It is a transcriptional repressor, expressed early in the anterior visceral endoderm and adjacent ectoderm in an area destined to become the ventral prosencephalon and forebrain. By E9.0, expression is restricted to the VD and presumptive RP [3]. Hesx1 appears to have a role in RP cellular proliferation and patterning, as its gradual down-regulation is concomitant with a gradual increase in the expression of another *paired-like* homebox gene, Prophet of Pit1 (*PROP1* – see below) [14]. This down-regulation of

Hesx1 is essential for pituitary development and expression ceases by E13.5 [3]. Initial activation of *Hesx1* is dependent upon *Lhx1* and *Lhx3* via its 5′ promoter region. 3′ elements are essential for later expression in the developing pouch [4].

In humans, *HESX1* mutations may occur in association with isolated growth hormone deficiency (IGHD) or combined pituitary hormone deficiency (CPHD), frequently in association with septo-optic dysplasia (SOD) phenotypes characterized by the presence of optic nerve hypoplasia or midline abnormalities of the brain. These patients exhibit variable MRI abnormalities ranging from a structurally normal pituitary to a more severe radiological phenotype characterized by anterior pituitary hypoplasia, an undescended or ectopic posterior pituitary, optic nerve hypoplasia and agenesis of the corpus callosum [2]. Mutations in *HESX1* can result in hypopituitarism without midline defects or optic nerve anomalies, as the pituitary is believed to be highly sensitive to Hesx1 dosage (table 1) [15].

SOX2 and SOX3

SOX2 and SOX3 are members of the SOX [SRY-related high mobility group (HMG) box] family of transcription factors and are early markers of progenitor cells; their expression is down-regulated as cells differentiate. Strong *SOX2* expression is detected within RP in human embryos (4.5–9 weeks of development), which is maintained throughout anterior pituitary development as well as in the overlying diencephalon. However, there is no *SOX2* expression in the infundibulum or posterior pituitary. Initially *SOX2* mutations had been associated with bilateral anophthalmia, severe microphthalmia, learning difficulties, oesophageal atresia and genital abnormalities (table 1). However, further phenotypic characterization has revealed the presence of anterior pituitary hypoplasia, hypogonadotrophic hypogonadism and variable GHD, often with accompanying phenotypes such as hippocampal abnormalities, corpus callosum agenesis, oesophageal atresia, hypothalamic hamartoma and sensorineural hearing loss [3]. Although in the majority of patients the MRI reveals a small anterior pituitary, in occasional cases, the pituitary is enlarged and remains so for years. The definitive Rathke's pouch comprises proliferative progenitors that will gradually relocate ventrally, away from the lumen as they differentiate. A proliferative zone containing progenitors is maintained in the embryo in a perilumenal area and was recently found to persist in the adult [4]. During pituitary development, SOX2 is expressed in the early ectoderm and maintained throughout the pouch. Its expression is down-regulated as endocrine cell differentiation proceeds. Expression is maintained in the prospective progenitor proliferative zone, around the Rathke's pouch lumen, during embryogenesis but also in the mature gland [4]. It is also expressed in the VD.

Normal hypothalamo-pituitary development is critically dependent upon the dosage of SOX3; over- or underdosage can lead to hypopituitarism or infundibular hypoplasia. *Sox3* null mutant mice have a variable phenotype showing poor growth, craniofacial defects, and variable endocrine deficits. In humans, SOX3 mutations are associated with variable phenotypes including either IGHD or panhypopituitarism.

There is variable developmental delay, and MRI usually reveals a small anterior pituitary, an ectopic/undescended posterior pituitary, and dysgenesis of the corpus callosum (table 1) [16]. Recently, a *SOX3* variant that leads to a deletion within a polyalanine tract has been associated with hypopituitarism in a heterozygous female patient; the variant has been shown to be associated with a gain in function as opposed to the previously described polyalanine expansions which led to loss of function [16, 17].

OTX2

Orthodentic homeobox 2 (*Otx2*) is a transcription factor whose role in hypothalamo-pituitary axial development remains largely unclear. Expression in mice is restricted to developing neural and sensory structures such as the brain, eyes, nose and ears. Recent elucidations show that *Otx2* is expressed in the VD by E10.5 where it may affect *Fgf8* and/or *Bmp* induction of RP formation [18]. Its expression at the same time in RP suggests an intrinsic role in RP development as well, which is consistent with its proposed ability to activate *Hesx1* expression. By E12.5, *Otx2* mRNA expression is undetectable in RP (however, it is still detectable in the VD) although the protein persists through to E14.5. By E16.5, Otx2/*Otx2* is absent from either RP or the VD. Homozygous-null mutant mice, which die midgestation, exhibit severe malformations of the forebrain structures as well as the eyes due to impaired gastrulation. Heterozygous mice present with variable eye phenotypes ranging from normal to severe (anophthalmia/microphthalmia) and brain structural abnormalities (holoprosencephaly or anencephaly). Its potential role in pituitary development was suggested in a recent report looking at *Prop1*-mutant mice [18]. In these animals, *Otx2* expression in RP persists through to E16.5, which is 4 days after peak expression of *Prop1* and 2 days later than any obvious pituitary morphology defects become apparent. This suggested that *Prop1* may regulate genetic expression of factors which may suppress *Otx2* expression and implies a role for *Otx2* in murine pituitary development. Further evidence of a role in hypothalamo-pituitary development was shown in GnRH-neuron-*Otx2* knockout mice which exhibited hypogonadotrophic hypogonadism [19]. These data are consistent with human *OTX2* phenotypes which can encompass highly variable hypopituitary phenotypes (ranging from isolated growth hormone deficiency to panhypopituitarism) and hypogonadotrophic hypogonadism, both of which are most commonly detected in association with severe eye abnormalities including anophthalmia or microphthalmia. To date, *OTX2* mutations, of which 20 have been described, account for 2–3% of severe eye abnormalities (table 1) [3].

Factors Regulating Cellular Differentiation
PROP1 and POU1F1/PIT1

Prop1 is a transcriptional activator whose primary role is to stimulate pituitary-cell differentiation. At E9.0, its expression is repressed by Hesx1. Subsequently, it is expressed in Rathke's pouch, and believed to form heterodimers with Hesx1. By E12.5, levels of Prop1 are believed to be sufficient to displace Hesx1 such that homodimers are formed. At this point, Prop1 stimulates the differentiation and proliferation of each of the hormone-secreting cell lineages. Mutations in *PROP1* are associated with GH, prolactin and TSH deficiencies, but in addition also manifest gonadotrophin and ACTH deficiencies which may be present at a young age, or may evolve with time. MRI reveals either a small or enlarged anterior pituitary that may wax and wane in size before eventual involution (table 1) [4, 20]. The posterior pituitary is structurally normal. Mutations in *PROP1* have in several studies been described as the most prevalent cause of combined pituitary hormone deficiency (CPHD) in humans, accounting for up to 50% of familial cases of CPHD. However, the prevalence of mutations is much lower in sporadic cases.

While the exact mechanisms of how Prop1 drives gonadotroph and corticotroph differentiation remain unclear, the stimulation of somatotroph, lactotroph and thyrotroph differentiation is driven via the activation of the POU domain, class 1, transcription factor (Pou1F1, also known as Pit1) at E13.5 [21]. Consequently, mutations in *POU1F1* are associated with GH, prolactin and TSH deficiencies in the presence of a normal or small anterior pituitary gland on MRI, with a structurally normal posterior pituitary (table 1). Another transcription factor, PITX1, is thought to synergise with POU1F1 to augment prolactin gene expression and, to a lesser extent, growth hormone gene expression [4]. In gonadotrophs it activates *Lhb* expression. Mutations in this gene have not been implicated in hypopituitarism in humans, however.

GATA2

GATA2 belongs to a family of six transcription factors. It has dual functions; (1) as a stem cell maintenance factor in some tissues, and (2) as a promoter of cellular differentiation in others [22]. Following induction by BMP2 at E10.5, *Gata2* is expressed in the ventral RP along with α-GSU, the common subunit of Lh, Fsh, and Tsh. It therefore marks the prospective, and then definitive, gonadotrophs and thyrotrophs; its expression is maintained in the adult. Gata2 can synergise with Pou1f1 to induce *Tshβ* expression and synergise with Nr5a1 to increase *Lhb* gene expression [4]. Despite its known role in regulating the human *α-GSU* gene, no mutations have been described in humans in association with hypopituitarism.

TBX19

TBX19 (previously referred to as TPIT) is a member of the T-Box family of transcription factors. *Tbx19* is exclusively expressed in the developing pituitary, first at E12.5 in *Pomc*-positive cells, then in corticotrophs and melanotrophs, where it is maintained in the adult gland [23]. *Pomc* transcription requires cooperation between Tbx19 and Pitx1, the two factors binding to contiguous sites within the same regulatory element. The essential role of Tbx19 for differentiation of pituitary Pomc lineages was ascertained in *Tbx19* null mutant mice which exhibit an almost complete loss of pituitary *Pomc*-expressing cells [4], resulting in severe Acth and glucocorticoid deficiencies. There is a mutual antagonism between Tbx19 and Nr5a1, the latter being important

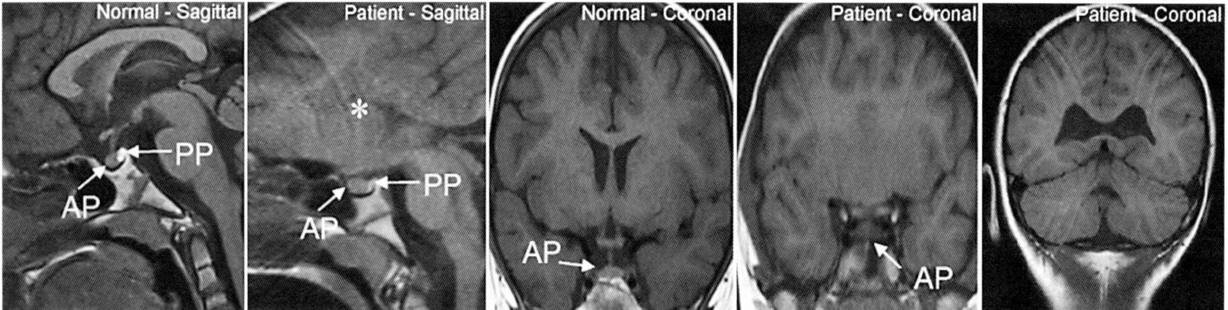

Fig. 3. Sagittal and coronal MRI scans of a normal control subject and a patient with the craniofacial/midline disorder – holoprosencephaly. The normal scan shows an intact corpus callosum above a healthy anterior pituitary (AP) and posterior pituitary (PP) as well as normal division of the brain into its cerebral hemispheres. In the patient, who has autosomal recessive holoprosencephaly due to a missense mutation in *FGF8*, the corpus callosum is absent (asterisk) and the anterior pituitary is enlarged. The failure of the brain to divide into its cerebral hemispheres is clearly evident in the coronal view. This Figure is an example and is not completely representative due to high variations in phenotypes seen across patients. The pituitary phenotype can include anterior pituitary hypoplasia and an ectopic posterior pituitary, neither of which are shown herein. Taken from McCabe et al., with permission.

for gonadotroph development. Therefore, Tbx19 promotes and is required for corticotroph and melanotroph fate determination while actively repressing gonadotroph differentiation. Mutations in *TBX19* are the commonest cause of isolated ACTH deficiency presenting with life-threatening hypoglycaemia in the neonatal period in humans (table 1).

Congenital Hypopituitarism and Associated Defects

As mentioned earlier, the pituitary gland is a midline structure and so defects affecting the development of this organ may impact on the development of associated structures, which can lead to highly heterogeneous phenotypes including HPE and SOD. Genetic mutations have been implicated in these disorders but in a very low proportion of cases (approximately 5–10% of patients studied) thereby indicating that many more causative genes or other aetiological factors remain to be uncovered.

HPE is the most common forebrain anomaly in humans, occurring at an incidence of 1:10,000–20,000 in the general population and as high as 1:250 in conceptuses [3]. It is caused by varying degrees of separation of the cerebral hemispheres and ventricles [3]. Failure of the frontal and parietal lobes to divide posteriorly result in the loss of the corpus callosum (fig. 3). Facial features associated with the disorder include cyclopia, anophthalmia, midface hypoplasia, hypotelorism, cleft lip and/or palate and a single central incisor. An increasing number of genetic factors and associated pathways have been implicated in the aetiology of HPE and include members of the Sonic Hedgehog pathway including SHH, ZIC2, TGIF1, PTCH1, GLI2, DISP1, TDGF1, GAS1, EYA4 and FOXH1 [10]. More recently, the first case of autosomal recessive HPE was assigned to a mutation in *FGF8* [10].

Septo-optic dysplasia (SOD) is diagnosed when any two of the following phenotypes are present: (1) optic nerve hypoplasia, (2) midline neuroradiological abnormalities such as agenesis of the corpus callosum and septum pellucidum, and (3) anterior pituitary hypoplasia with consequent hypopituitarism [3]. While generally sporadic and inclusive of possible environmental induced pathologies through smoking and drug/alcohol abuse, there is an increasing number of implicated genetic mutations involving *HESX1, OTX2, SOX2* and *SOX3* (table 1). As mentioned earlier, these genes, and those implicated in HPE, are expressed early in forebrain and pituitary development and, as such, mutations within may be expected to induce congenital hypopituitarism and associated phenotypes.

Overlap between Congenital Hypopituitarism and Midline Defects with Kallmann Syndrome

Recent evidence has suggested a possible overlap between Kallmann syndrome and midline disorders that include hypopituitarism. A number of phenotypic features are shared by the two disorders including craniofacial defects (high-arched palate, cleft lip/palate) [3], dental agenesis and sensorineural hearing loss. A number of genes have been implicated in the aetiology of Kallmann syndrome, including *KAL1, FGFR1, FGF8, PROKR2* and *PROK2* with recent additions including the chromodomain helicase DNA binding protein-7 (CHD7), Bromodomain and WD repeat-containing protein 2 (WDR11), negative elongation factor (NELF) and Semaphorin 3A (SEMA3A) [24–26].

The original Kallmann syndrome associated genes are involved in the maturation and/or migration of gonadotrophin-releasing hormone neurons from the olfactory placode to the hypothalamus. These genes are also involved in the maturation of the olfactory complex itself, hence the anosmic phenotype characteristic of the condition. Recently, mutations in *FGFR1* and *FGF8* have been implicated in CPHD, SOD and HPE highlighting an overlap in genotypes between these conditions and consistent with the overlap in phenotypes [10, 24]. Furthermore, *PROKR2* sequence variants that had previously been implicated in the aetiology of KS have also been recently identified in patients with congenital hypopituitarism and SOD [24]. Thus, as previously observed with *HESX1*, sequence variants may be associated with variably penetrant phenotypes. This is suggestive of a possible digenic or oligogenic cause of congenital hypopituitarism; a phenomenon which is now well established in the aetiology of KS [24]. Thus, the molecular basis of these disorders is complex and could involve defects in multiple factors, as well as possible interactions with the environment.

Current data suggest that the aetiological basis of congenital hypopituitarism is far from straightforward; given that only between 5 and 15% of cases have an associated genetic variant. The advent of homozygosity mapping and whole exome sequencing however opens new avenues for the identification of novel genes implicated in congenital hypopituitarism and associated disorders.

Conclusions

The development of the pituitary is a highly complex and organised process involving spatio-temporally arranged signalling molecules and transcription factors. This chapter has provided an overview of pituitary development and some of the molecules involved in the process. Mutations in the genes encoding these molecules can lead to congenital hypopituitarism, including disorders such as HPE and SOD. Such genetic anomalies only account for a small percentage of cases although the apparent overlap in genotypes with Kallmann syndrome opens a new, albeit limited, source of potential genetic associations. It is expected that with the advent of homozygosity mapping and whole exome sequencing, new genes will be implicated in both of these groups of disorders thereby enhancing our understanding of pituitary development and how defects in these newly identified genes may impact on human phenotypes.

References

1 Cohen LE: Genetic disorders of the pituitary. Curr Opin Endocrinol Diabetes Obes 2012;19:33–39.

2 Alatzoglou KS, Dattani MT: Genetic forms of hypopituitarism and their manifestation in the neonatal period. Early Hum Dev 2009;85:705–712.

3 McCabe MJ, Alatzoglou KS, Dattani MT: Septo-optic dysplasia and other midline defects: the role of transcription factors: *HESX1* and beyond. Best Pract Res Cl En 2011;25:115–124.

4 Kelberman D, Rizzoti K, Lovell-Badge R, Robinson ICAF, Dattani MT: Genetic regulation of pituitary gland development in human and mouse. Endocr Rev 2009;30:790–829.

5 Reis LM, Tyler RC, Schilter KF, Abdul-Rahman O, Innis JW, Kozel BA, Schneider AS, Bardakjian TM, Lose EJ, Martin DM, Broeckel U, Semina EV: *BMP4* loss-of-function mutations in developmental eye disorders including SHORT syndrome. Hum Genet 2011;130:495–504.

6 Bakrania P, Efthymiou M, Klein JC, Salt A, Bunyan DJ, Wyatt A, Ponting CP, Martin A, Williams S, Lindley V, Gilmore J, Restori M, Robson AG, Neveu MM, Holder GE, Collin JRO, Robinson DO, Farndon P, Johansen-Berg H, Gerrelli D, Ragge NK: Mutations in *BMP4* cause eye, brain, and digit developmental anomalies: Overlap between the *BMP4* and hedgehog signalling pathways. Am J Hum Genet 2008;82:304–319.

7 Traiffort E, Dubourg C, Faure H, Rognan D, Oden S, Durou MR, David V, Ruat M: Functional characterization of sonic hedgehog mutations associated with holoprosencephaly. J Biol Chem 2004;279:42889–42897.

8 França MM, Jorge AAL, Carvalho LRS, Costalonga EF, Vasques GA, Leite CC, Mendonca BB, Arnhold IJP: Novel heterozygous nonsense GLI2 mutations in patients with hypopituitarism and ectopic posterior pituitary lobe without holoprosencephaly. J Clin Endocr Metab 2010;95:E384–E391.

9   Zhao M, Qiao M, Harris SE, Chen D, Oyajobi BO, Mundy GR: The zinc finger transcription factor Gli2 mediates bone morphogenetic protein 2 expression in osteoblasts in response to hedgehog signalling. Mol Cell Biol 2006;26:6197–6208.

10 McCabe MJ, Gaston-Massuet C, Tziaferi V, Gregory LC, Alatzoglou KS, Signore M, Puelles E, Gerrelli D, Farooqi IS, Raza J, Walker J, Kavanaugh SI, Tsai PS, Pitteloud N, Martinez-Barbera JP, Dattani MT: Novel *FGF8* mutations associated with recessive holoprosencephaly, craniofacial defects and hypothalamo-pituitary dysfunction. J Clin Endocr Metab 2011;96:E1709–E1718.

11 Sheng HZ, Zhadanov AB, Mosinger B Jr, Fujii T, Bertuzzi S, Grinberg A, Lee EJ, Huang SP, Mahon KA, Westphal H: Specification of pituitary cell lineages by the LIM homeobox gene *Lhx3*. Science 1996;272:1004–1007.

12 Sheng HZ, Moriyama K, Yamashita T, Li H, Potter SS, Mahon KA, Westphal H: Multistep control of pituitary organogenesis. Science 1997;278:1809–1812.

13 Desmaison A, Vigouroux A, Rieubland C, Peres C, Calvas P, Chassain N: Mutations in the *LHX2* gene are not a frequent cause of micro/anophthalmia. Mol Vis 2010;16:2847–2849.

14 Carvalho LR, Brinkmeier ML, Castinetti F, Ellsworth BS, Camper SA: Corepressors TLE1 and TLE3 interact with HESX1 and PROP1. Mol Endocrinol 2010;24:754–765.

15 Sajedi E, Gaston-Massuet C, Signore M, Andoniadou CL, Kelberman D, Castro S, Etchevers HC, Gerrelli D, Dattani MT, Martinez-Barbera JP: Analysis of mouse models carrying the I26T and R160C substitutions in the transcriptional repressor HESX1 as models for septo-optic dysplasia and hypopituitarism. Dis Model Mech 2008;1:241–254.

16 Alatzoglou KS, Kelberman D, Cowell CT, Palmer R, Arnhold IJ, Melo ME, Schnabel D, Grueters A, Dattani MT: Increased transactivation associated with SOX3 polyalanine tract deletion in a patient with hypopituitarism. J Clin Endocrinol Metab 2011;96:E658–E690.

17 Woods KS, Cundall M, Turton J, Rizotti K, Mehta A, Palmer R, Wong J, Chong WK, Al-Zyoud M, El-Ali M, Otonkoski T, Martinez-Barbera JP, Thomas PQ, Robinson IC, Lovell-Badge R, Woodward KJ, Dattani MT: Over- and underdosage of SOX3 is associated with infundibular hypoplasia and hypopituitarism. Am J Hum Genet 2005;76:833–849.

18 Mortensen AH, MacDonald JW, Gosh D, Camper SA: Candidate genes for panhypopituitarism identified by gene expression profiling. Physiol Genomics 2011;43:1105–1116.

19 Diaczok D, Divall S, Matuso I, Wondisford FE, Wolfe AM, Radovick S: Deletion of Otx2 in GnRH neurons results in a mouse model of hypogonadotrophic hypogonadism. Mol Endocrinol 2011;25:833–846.

20 Ward RD, Raetzman LT, Suh H, Stone BM, Nasonkin IO, Camper SA: Role of PROP1 in pituitary gland growth. Mol Endocrinol 2005;19:698–710.

21 Bodner M, Castrillo JL, Theill LE, Deerinck T, Ellisman M, Karin M: The pituitary specific transcription factor GHF-1 is a homeobox-containing protein. Cell 1988;55:505–518.

22 Morceau F, Schnekenburger M, Dicato M, Diederich M: GATA-1: friends, brothers, and coworkers. Ann NY Acad Sci 2004;1030:537–554.

23 Lamolet B, Pulichino AM, Lamonerie T, Gauthier Y, Brue T, Enjalbert A, Drouin J: A pituitary cell-restricted T box factor, Tpit, activates POMC transcription in cooperation with Pitx homeoproteins. Cell 2001;104:849–859.

24 Raivio T, Avbelj M, McCabe MJ, Romero CJ, Dwyer AA, Tommiska J, Sykiotis GP, Gregory LC, Diaczok D, Tziaferi V, Elting MW, Padidela R, Plummer L, Martin C, Feng B, Zhang C, Zhou QY, Chen H, Mohammadi M, Quinton R, Sidis Y, Radovick S, Dattani MT, Pitteloud N: Genetic overlap in Kallmann syndrome, combined pituitary hormone deficiency and septo-optic dysplasia. J Clin Endocrinol Metab 2012;97:E694–E699.

25 Kim HG, Layman LC: The role of CHD7 and the newly identified WDR11 gene in patients with idiopathic hypogonadotrophic hypogonadism and Kallmann syndrome. Mol Cell Endocrinol 2011;346:74–83.

26 Young J, Metay C, Bouligand J, Tou B, Francou B, Maione L, Tosca L, Sarfati J, Brioude F, Esteva B, Briand-Suleau A, Brisset S, Goossens M, Tachdjian G, Guiochon-Mantel A: *SEMA3A* deletion in a family with Kallmann syndrome validates the role of semaphorin 3A in human puberty and olfactory system development. Hum Reprod 2012;27:1460–1465.

Prof. Mehul T. Dattani
UCL-Institute of Child Health
30 Guilford Street
London, WC1N 1EH (UK)
Tel. +44 207 905 2657, E-Mail m.dattani@ucl.ac.uk

Mullis P-E (ed): Developmental Biology of GH Secretion, Growth and Treatment.
Endocr Dev. Basel, Karger, 2012, vol 23, pp 16–29 (DOI: 10.1159/000341736)

Pituitary Gland Imaging and Outcome

Natascia Di Iorgi · Giovanni Morana · Anna Lisa Gallizia ·
Mohamad Maghnie

Departments of Pediatrics and Neuroradiology (GR), IRCCS Giannina Gaslini Institute, University of Genova,
Genova, Italy

Abstract

Magnetic resonance imaging (MRI) allows a detailed and precise anatomical study of the pituitary
gland by differentiating between the anterior and posterior pituitary lobes. The identification of
posterior pituitary hyperintensity, now considered a marker of neurohypophyseal functional
integrity, has been the most striking advance for the diagnosis and understanding of anterior and
posterior pituitary diseases. The advent of MRI has in fact led to a significant improvement in the
understanding of the pathogenesis of disorders that affect the hypothalamo-pituitary area. Today,
there is convincing evidence to support the hypothesis that marked MRI differences in pituitary
morphology indicate a diverse range of disorders which affect the organogenesis and function of
the anterior pituitary gland with different prognoses. Furthermore, the association of extrapitu-
itary malformations accurately defined by MRI has supported a better definition of several conditions linked to pituitary hormone deficiencies and midline defects. MRI is a very informative
procedure that should be used to support a diagnosis of hypopituitarism. It is useful in clinical
management, because it helps endocrinologists determine which patients to target for further
molecular studies and genetic counselling, which ones to screen for additional hormone deficits,
and which ones may need growth hormone replacement into adult life.

Magnetic resonance imaging (MRI) is the radiological examination method of choice
for evaluating hypothalamo-pituitary-related endocrine diseases and is considered
essential in the assessment of patients with suspected pituitary involvement [1, 2].
The use of MRI has led to an enormous increase in our comprehensive knowledge
of pituitary morphology, improving the differential diagnosis of hypopituitarism in
particular. Specifically, MRI allows for a detailed and precise anatomical study of the
pituitary gland by differentiating between the anterior and posterior pituitary lobes.
MRI recognition of pituitary hyperintensity in the posterior part of the *sella*, now considered a marker of neurohypophyseal functional integrity, has been the most striking
development for the diagnosis and understanding of certain forms of 'idiopathic' and
permanent growth hormone deficiency (GHD) [3–14]. However, these developments

have inevitably raised some important questions. Is there a gold standard for the diagnosis of GHD and what then is the ultimate goal of MRI examination? Is MRI phenotype imaging informative? Does a diagnosis of MRI phenotype affect the work-up of patients with GHD? Can MRI phenotype predict the outcome of patients with GHD? And, finally, what is the contribution of these MRI findings to overall GHD management? Thus, MRI is certainly an important benchmark, but several questions remain unanswered and others clearly represent a challenge. This article reviews the neuroimaging characteristics and impact of MRI findings in the current management and outcome of diseases involving the hypothalamic-pituitary area.

Imaging of Normal Pituitary

The correct interpretation of MRI scans requires a detailed knowledge of the normal features of the pediatric pituitary gland and of the changes it goes through in the same individual over time.

The assessment of the pituitary gland includes the evaluation of signal intensity, shape, size, position and connection with surrounding tissues. The evaluation of central nervous system (CNS) structures including the corpus callosum, septum pellucidum, optic chiasm, olfactory tracts and bulb, cerebellum, and pons is mandatory.

Normal Pituitary Size. Pediatric pituitary gland size and shape change physiologically throughout life depending on age and gender. Moreover, even among children of identical age and gender there is a large degree of morphological and dimensional variability [15–17]; the morphology of the gland may also vary from side to side reflecting a physiological asymmetric development of the pituitary fossa, which in turn may be influenced by a varying degree of pneumatization of the underlying sphenoid sinus throughout infancy. With this in mind, a simple measure of a single dimension (i.e. height) cannot be considered a perfectly reliable indicator of the size of a tridimensional structure such as the pituitary gland. Nevertheless, measurement of pituitary gland height is still the most widely used method to obtain a rapid, indirect determination of gland size. The maximum height of the anterior pituitary is measured drawing a perpendicular line parallel to the hard palate and tangent to the sella turcica floor [18]. Normal pituitary gland height values range between 3 and 6 mm in prepubertal children. In particular, in newborns, the gland typically presents an upward convex margin with a mean height (about 4 mm) that is slightly higher when compared to the gland height in the following months of age. In general, the height of the pituitary gland either slightly decreases or remains stable during the first 2 years of life, whereas its width (transverse measurement) and depth (anterior-posterior measurement) slightly increase; subsequently, a mild but progressive increase in height follows until puberty, when the pituitary gland undergoes rapid and profound changes in size and shape, with marked enlargement. In girls, the gland may swell symmetrically to a height of 10–12 mm, appearing nearly spherical, whereas in pubertal boys it may reach 7–8 mm

[1, 2, 15, 19]. Indirect pituitary volume assessment has been calculated using formulas adapted from the formula for the volume of an ellipsoid (i.e. the Di Chiro formula: V = 1/2 length × height × width). Since glands are usually not spherical, these methods do not allow a precise calculation of the real size of the gland [20]. In overcoming the above-cited computational bias, volumetric data, based on a direct calculation of pituitary volumes from 3D MRI sequences, are a significant improvement over the information collected using earlier methods. A 3D data set of T_1-weighted images can be reformatted on sagittal or coronal planes and the entire volume is calculated adding the volume of the gland on each slice [20–22]. Unfortunately, this method is time consuming and of limited use in daily practice. Furthermore, a careful comparison of the results obtained by three different [20–22] prior studies shows a relative discrepancy among results, with wide variations of normal ranges. For example, in one study [20] where the volume of the gland included the posterior pituitary, the average volume of the entire gland resulted smaller at a given age, when compared to a different study where only the volume of the adenohypophysis was taken into account [21]. A better standardization of the computation method and larger study series are awaited to better elucidate the clinical significance and reproducibility of volumetric results that, at present, appear somewhat incomplete.

Empty Sella. The so-called 'empty sella' is a misnomer since the sella is almost never truly empty [23]. The term was introduced in the pre-CT, pre-MRI era, when neurosurgeons used to operate a suspected pituitary mass lesion on the basis of an expanded sella turcica on plain radiography. In many cases, the sella was then found to be filled only with cerobrospinal fluid (CSF) instead of a tumor, with the pituitary gland flattened along the sellar floor. At present, the term empty sella basically indicates an intrasellar herniation of the subarachnoid spaces due to an incompetent sellar diaphragm (arachnoidal diverticulum) [19, 24]. Rhythmic CSF pulsations contribute to chronic compression of the pituitary gland and enlargement of the pituitary fossa (so-called primary empty sella). MRI clearly demonstrates the presence of a deep and enlarged pituitary fossa mainly filled with CSF, containing an anterior pituitary lobe that appears as a thin layer along its floor. The posterior lobe is flattened against the dorsum sellae, with a thin and stretched pituitary stalk. An intrasellar arachnoid expansion can also occur secondary to reduced intrasellar tissue volume due to such causes as surgical resection, radiation necrosis, pituitary atrophy, or infarction (secondary empty sella). In such cases, it is essentially an ex vacuo phenomenon, where intracranial subarachnoid space secondarily extends into the sella. Previous studies have reported the prevalence of empty sellae to be about 5–9% for all ages, with an increase in prevalence with age [25, 26]. Most frequently, this anomaly is an incidental finding of little or no clinical importance, the only exception being pseudotumor cerebri (benign intracranial hypertension) in which clinical symptoms and additional MRI findings may orient the diagnosis [27]. Semantic problems can also arise with the so-called 'partially empty sella', another misnomer, in which only part of the intrasellar space is filled with an arachnoid diverticulum; in this case, the height of the gland is reduced on the midline, but gland tissue is seen to extend to full height along

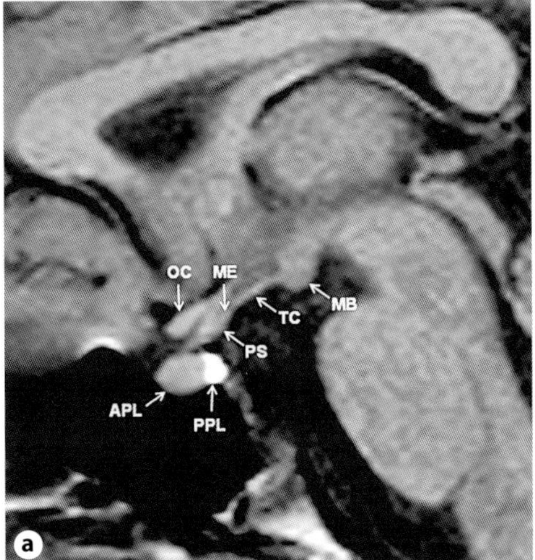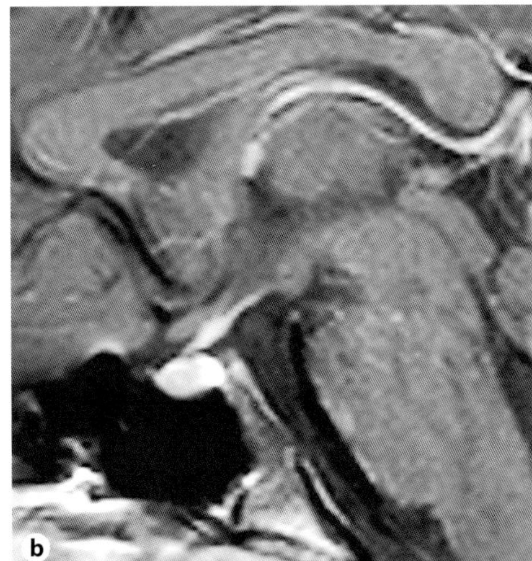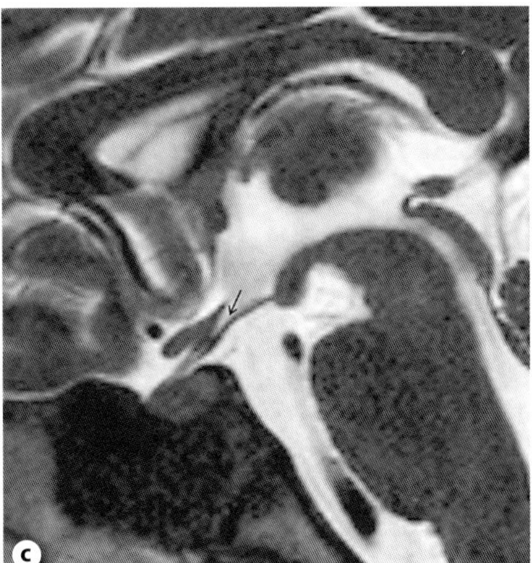

Fig. 1. Normal sellar region. **a** Sagittal T_1-weighted image. **b** Gd-enhanced sagittal T_1-weighted image. **c** Sagittal T_2-driven equilibrium (DRIVE) sequence obtained at 0.7 mm thickness. **a** The posterior pituitary lobe (PPL), physiologically bright on T_1-weighted image, anterior pituitary lobe (APL), pituitary stalk (PS), median eminence (ME), optic chiasm (OC), tuber cinereum (TC), and mammillary bodies (MB) are clearly visible. Following gadolinium injection (**b**), there is homogeneous enhancement of the pituitary gland, pituitary stalk, median eminence, and tuber cinereum. High-resolution, heavily T_2-weighted sequence (**c**) allows an optimal depiction of the pituitary stalk with sharp delineation of the median eminence and infundibular recess of the third ventricle (black arrow).

the bilateral sellar margins. This condition is not easy to distinguish from a normal or hypoplastic pituitary gland on MR images. The presence of a small pituitary fossa may help in distinguishing pituitary hypoplasia from a partially empty sella.

High-Resolution, Heavily T_2-Weighted Images. High-resolution, heavily T_2-weighted images (i.e. T_2-DRIVE, constructive interference in steady state or fast imaging employing steady-state acquisition sequence) obtained at submillimetric thickness allow an excellent and detailed representation of the anatomy of the suprasellar compartment, and in particular of the pituitary stalk; pituitary stalk thickness is better evaluated with this technique than with conventional precontrast T_1- and T_2-weighted images and, in our experience, its sensitivity is similar to that of post contrast T_1-weighted images (fig. 1). These 3D sequences also allow a better evaluation of the pituitary stalk on follow-up studies, as these images may be reformatted into orthogonal planes in order to obtain triplanar information and an identical slice orientation when compared to prior studies. Intrasellar or parasellar cysts are also extremely well depicted, as well as adjacent midline structures, including the median eminence, tuber cinereum and mammillary bodies; the main limitation of these sequences is the homogeneous low signal intensity of the pituitary gland that cannot be confidently separated into the anterior and posterior lobe. We ultimately recommend the routine use of these sequences, particularly when a serial and accurate evaluation of pituitary stalk thickness is required and/or when a detailed anatomical representation of the suprasellar region is needed, especially for surgical purposes.

MRI Phenotypes and Outcome

Idiopathic GH Deficiency and Structural Hypothalamo-Pituitary Abnormalities. A recent study analyzed the clinical phenotype in a group of 137 GHD-patient photographs that contained 73 subjects with IGHD and 64 with multiple pituitary hormone deficiencies (MPHD). Standardized frontal and lateral digital pictures were taken of each patient and analyzed using dedicated software. The analysis revealed that Canthal Index (CI), the relative distance between the eyes, was correlated with pituitary morphology. Patients with an ectopic posterior pituitary (EPP) had significantly higher CI values than patients without EPP, with a cut-off value of CI >39 for identifying children with the highest probability of having EPP. The combination of CI >39 together with the presence of hormonal deficiencies in addition to GHD strongly predicted EPP: 93% of the patients with a CI >39 and additional hormonal deficiencies had EPP, compared to only 77% of patients with additional hormonal deficiencies who had a CI <39. Only 29% of the patients without either CI >39 or additional hormonal deficiencies had EEP. The association between Canthal Index, measured from digital pictures, and EPP could be due to an altered midline development that affects both the pituitary gland and the facial structures of GHD patients [28].

MRI Evaluation of Hypothalamic Pituitary Region and CNS Abnormalities. The MRI findings include (a) normal or hypoplastic pituitary gland without anatomical abnormalities of the hypothalamus or pituitary stalk (fig. 2), and (b) severe-to-moderate (2–3 mm) hypoplastic pituitary gland with ectopic posterior pituitary located anywhere from the median eminence to the distal stalk [1–18] (fig. 3–5). Isolated GHD is more commonly observed in the former category whereas MPHD occurs more frequently in the latter. MPHD may be associated with a spectrum of cerebral malformations (fig. 6) such as Arnold-Chiari I and II, agenesis of the septum pellucidum, septo-optic dysplasia, vermis dysplasia, syringomyelia, absence of the internal carotid artery, dysgenesis of the corpus callosum, arachnoid cysts, and tentorium anomalies with basilar impression [2, 7]. The frequency of these radiological findings and their spectrum of pituitary

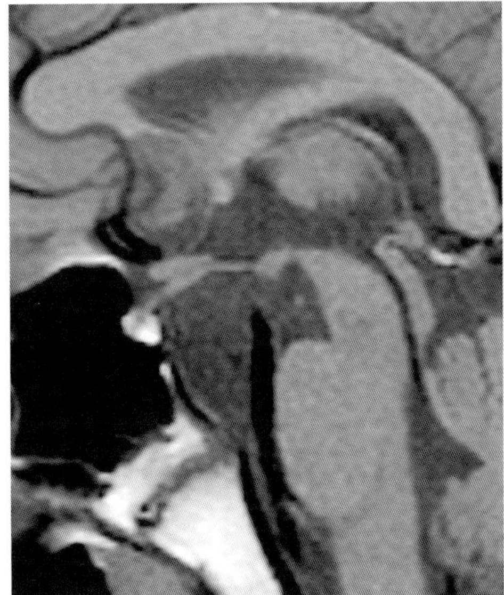

Fig. 2. Anterior pituitary hypoplasia in a 12-year-old boy with isolated GH deficit. Sagittal T_1-weighted image. Small anterior pituitary lobe housed within a small pituitary fossa. Normal posterior lobe and pituitary stalk.

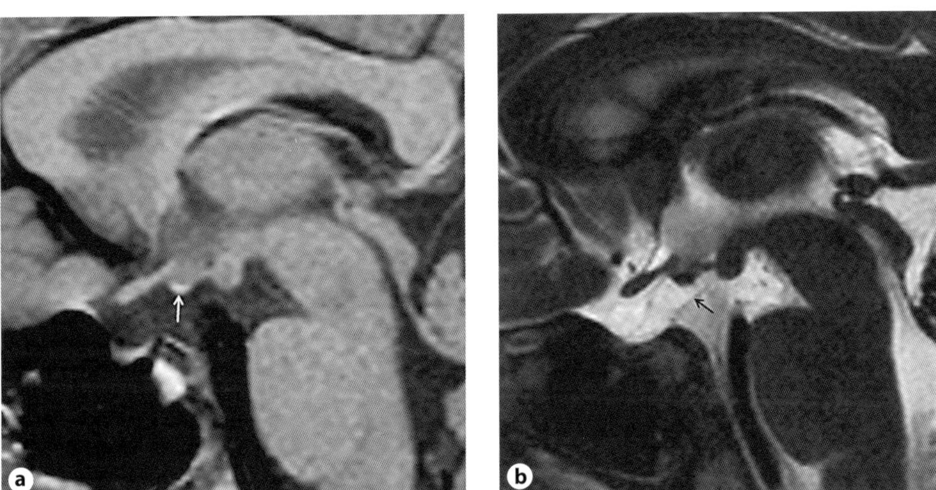

Fig. 3. Pituitary dystopia. **a** Sagittal T_1-weighted image. **b** Sagittal T_2-DRIVE image. Hypoplastic anterior lobe. Small and ectopic posterior lobe, located at the level of the median eminence (white arrow, **a**). The pituitary stalk is not visible both on T_1- and T_2-DRIVE images. Of note, high-resolution T_2-DRIVE image shows the Liliequist membrane (black arrow, **b**), not visible with conventional T_1- and T_2-weighted images.

hormone deficiencies are variable. In particular, MPHD is more frequently associated than isolated GHD with ectopic posterior pituitary and with anterior pituitary hypoplasia [1–18]. The variability between different studies can be attributed, variously, to the degree of restriction in the studies' diagnostic criteria, to the diagnostic limits of GHD itself (transitory deficits, recovery, false positives, etc.), and/or to the lack of

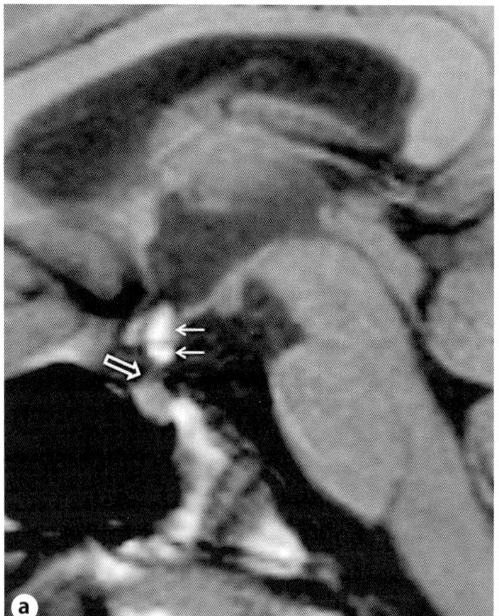 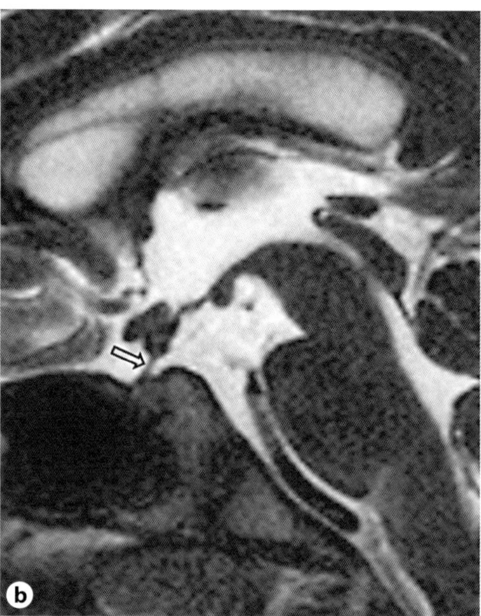

Fig. 4. Pituitary dystopia. **a** Sagittal T$_1$-weighted image. **b** Sagittal T$_2$-DRIVE image. There is a bulky ectopic posterior lobe with bilobated appearance, located at the level of the median eminence and within the proximal third of the infundibulum (thin arrows, **a**). The pituitary stalk is barely recognizable on T$_1$-weighted image (open white arrow, **a**) whereas is confidently demonstrated on T$_2$-DRIVE image (open black arrow, **b**).

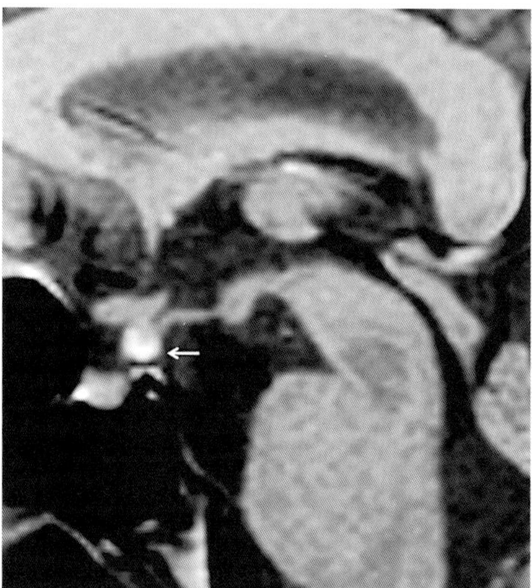

Fig. 5. Pituitary dystopia. Sagittal T$_1$-weighted image. Ectopic posterior lobe located at the middle third of the pituitary stalk (arrow). The distal portion of the infundibulum is not recognizable.

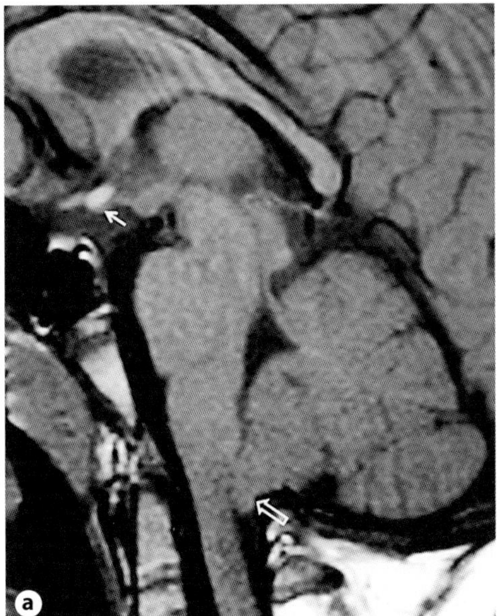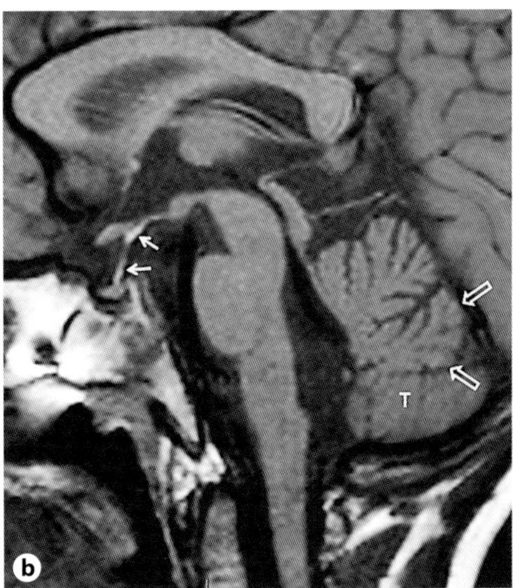

Fig. 6. Pituitary dystopia and associated malformations. **a** Sagittal T_1-weighted image. Hypoplastic anterior lobe housed within a small pituitary fossa and ectopic posterior lobe located at the level of the median eminence (thin arrow). The pituitary stalk is not visible. There is also downward extension of the cerebellar tonsils into the foramen magnum in keeping with a mild form of Chiari I malformation (open arrow). **b** Sagittal T_1-weighted image. Hypoplastic anterior lobe. The posterior lobe is ectopic and located at the level of the median eminence and along the distal third of the infundibulum (thin arrows). There is also a dysplastic vermis with inferior vermian hypoplasia (open arrows). T = Cerebellar tonsil.

a convincing standard for a normal size pituitary gland among the pre-pubertal pediatric population. Idiopathic MPHD has been reported less frequently in association with anterior pituitary hypoplasia and more frequently with normal posterior pituitary and pituitary stalk; the high frequency of sporadic forms of idiopathic GHD associated with ectopic posterior pituitary in the absence of genetic identification remains intriguing, suggesting that other factors may play a role as well.

The identification of EPP or pituitary stalk agenesis itself is helpful in the diagnosis and prognosis of patients with GH deficiency (and the size of EPP may vary considerably between patients, ranging from small EPP to large EPP and even huge EPP). In particular, the *size* and the *location* of EPP are early markers of evolving pituitary hormone deficiencies as reported in studies showing that small EPP surface area and/or hypothalamic-sited EPP are predictive of the development of MPHD [29–30]. On the other hand, it is worth pointing out that a hypothalamic location of EPP is generally associated with complete pituitary stalk agenesis and that a huge EPP was misdiagnosed as a subhypothalamic tumor in a patient subsequently subjected to unnecessary neurosurgery [31].

Table 1. MRI features and risk of pituitary dysfunction

Phenotype	High risk	Intermediate risk	Low risk
Clinical	CI >39	–	–
Anterior pituitary size	height <2 mm	2–3 mm	>3 mm
Pituitary stalk	agenesis	hypoplasia	normal
Posterior pituitary size	XS	small	normal
Posterior pituitary site	median eminence	along the pituitary stalk	normal
CNS abnormalities	complex phenotype	minor abnormalities	absent
Anterior pituitary dysfunction	MPHD	CPHD/IGHD	IGHD
Posterior pituitary dysfunction	complex CNS abnormalities and EPP	minor CNS abnormalities	–

CNS = Central nervous system; XS = extra small; MPHD = multiple pituitary hormone deficiencies; CPHD = combined pituitary hormone deficiencies; IGHD = isolated GHD; EPP = ectopic posterior pituitary.

MRI and Evolving Pituitary Dysfunction. The presence of a vascular component of the stalk has prognostic significance since patients with agenesis of the pituitary stalk run a greater risk of developing multiple pituitary hormone deficits than those who show a vascular residue of the stalk. Patients in whom a pituitary stalk cannot be identified after gadolinium (Gd-DTPA) administration have a risk of developing MPHD evolving to MPHD that is 27 times greater than those with a residual vascular pituitary stalk [9, 32].

A detailed study of the pituitary stalk with administration of Gd-DTPA is no longer recommended provided that T2-DRIVE has been performed. T2-DRIVE obtained at submillimetric thickness allows for an excellent and detailed representation of the anatomy of the suprasellar compartment, and in particular of the pituitary stalk; pituitary stalk thickness is better evaluated than with conventional precontrast T_1- and T_2-weighted images and, in our experience, the sensitivity is similar to that of post contrast T_1-weighted images.

MRI and Height Response to rhGH Treatment. Evidence of congenital developmental abnormalities at MR imaging is a stronger predictor than the maximal stimulated GH response of height gain during the first 3 years of rhGH treatment in prepubertal children with GHD [33]. In addition, whereas patients with normal pituitary size who have been treated with rhGH do not show any difference in catch-up growth or in achievement of adult height compared to untreated subjects with idiopathic short stature, significantly higher final stature with a greater height gain was, however, observed in patients with pituitary gland abnormalities at MRI [1, 34]. The presence of gonadotropin deficiency does not significantly influence statural prognosis in terms of adult height and pubertal height gain, as recently demonstrated in two cohorts of subjects with childhood onset IGHD or MPHD associated with structural hypothalamo-pituitary abnormalities [12].

MRI Findings and Pituitary Genes. Our understanding of the genetic regulation of pituitary gland development in humans and the mouse has also increased considerably, and mutations in a number of genes have been associated with pituitary dysfunction and abnormal pituitary gland development.

Animal and human studies, along with the correlation of a particular genetic profile to certain endocrine and MRI phenotypes, have yielded great insights into pituitary development [2].

Implications of MRI Pituitary Morphology after Adult Height Achievement in the Definition of Transient or Permanent GHD. The diagnosis of GHD in young adults is not straightforward and represents a major clinical challenge. The key predictors of persistent GHD are the severity of the original GH deficiency, the presence of additional pituitary hormone deficits, severely low IGF-1 concentrations, and structural HP abnormalities [13, 35]. We have demonstrated that patients with GHD and congenital hypothalamo-pituitary abnormalities may not require re-evaluation of GH secretion, whereas patients with isolated GHD and normal or small pituitary gland should be retested well before the attainment of adult height [13]. MRI findings of the hypothalamo-pituitary area in patients with GHD may be the most important criterion upon which the decision to re-evaluate a patient can be based, as opposed to response to pharmacological stimulation.

These observations are in line with another recent study evaluating GH secretion after arginine stimulation test in two cohorts of subjects with childhood onset GHD and ectopic or normally located posterior pituitary (EPP vs. NPP) at the attainment of adult height. The study demonstrates that the presence of an EPP compared to an NPP increases the likelihood of persistent GHD by 26% [14].

Two recent Consensus Statements on the management of young adults with childhood onset GHD during transition phase [36, 37] state that there are various groups of patients that have a high likelihood of permanent GHD after adult height attainment; these include those with severe GHD in childhood with or without two or three additional hormone deficits possibly due to a defined genetic cause, those with severe GHD due to structural hypothalamic-pituitary abnormalities and with CNS tumors, and patients who have received high-dose cranial irradiation.

A subgroup of subjects with idiopathic GHD of childhood onset presenting with congenital structural hypothalamo-pituitary abnormalities confirms that GHD patients – defined a priori as those with GH response <5 µg/l and with anterior pituitary hypoplasia, pituitary stalk agenesis and posterior pituitary ectopia at the level of the median eminence – are likely candidates for permanent GHD in adult life [38], while those with less severe MRI features have an uncertain diagnosis or a likelihood of normal GH response after stimulation tests. These findings have important clinical implications in the diagnosis and prognosis of GHD after adult height achievement. By applying the criterion of peak GH values of less than 3 or 5 µg/l, several misdiagnosed GHD subjects would be wrongly excluded from a potentially beneficial renewal of GH replacement treatment [39]. The comparison between a cohort of subjects with a high likelihood of permanent GHD (defined as those with hypothalamo-pituitary abnormalities at MRI) and a control group showed that the lowest values observed in normal subjects of peak GH after insulin tolerance test (ITT) of 6.1 µg/l and IGF-1 of –1.7 SDS could identify GHD subjects with a sensitivity of 96% and 77%, respectively, and a specificity of 100% for both groups [39]. Another recent study that compared two groups of subjects with high and low likelihood of permanent GHD confirmed that a peak GH response after insulin tolerance test of less than 5.62 µg/l can distinguish between the two groups, correctly classifying 87.34% of subjects (with a sensitivity of 77.42% and specificity of 93.75%). Indeed, IGF-1 measurement is a useful marker of the degree of GHD, ensuring an accurate classification of patients with severe GHD at cut-off levels of ≤–2.8 SDS [40].

Pituitary function should be periodically assessed in subjects with pituitary stalk agenesis and IGHD or CPHD, as they may develop additional pituitary hormone deficiencies and deterioration of metabolic parameters [41]. In our recent study, ACTH deficiency characterized a subset of patients with idiopathic GHD and pituitary stalk abnormalities, revealing that several of them had undiagnosed subclinical ACTH deficiency [42]. In the presence of thirst and water unbalance, recent data suggest that the most common posterior pituitary dysfunction in septo-optic dysplasia was identified as hyperosmolality and relative AVP deficiency characterizing patients with complex brain abnormalities and ectopic posterior pituitary. Indeed, patients with EPP and congenital hypopituitarism have normal posterior pituitary function [43].

Conclusions

Establishing endocrine and MRI phenotypes is extremely helpful in the identification and management of patients with hypopituitarism, both in terms of possible etiological definition and for the early diagnosis of evolving anterior pituitary hormone deficiencies. The first-line identification of EPP during childhood is predictive of anterior pituitary dysfunction which allows for the selection of patients for the molecular analyses of genes involved in GH secretion or pituitary cell differentiation (GH-N, GHRHR, POUF1F1, PROP1, LHX3, others). Neuroimaging is expected to progressively expand and improve our knowledge and understanding of pituitary diseases.

Key Issues

- MRI imaging is the technique of choice in the diagnosis of children with hypopituitarism.
- Marked differences in MRI pituitary gland morphology suggest different etiologies of GHD and different prognoses.

- Pituitary stalk agenesis and ectopic posterior pituitary are specific markers of permanent GHD and show a different clinical and endocrine outcome compared to those with hypoplastic pituitary stalk or normal pituitary stalk anatomy.
- T2 DRIVE aids in the identification of pituitary stalk without the use of contrast medium administration.
- Future developments in imaging techniques will undoubtedly reveal additional insights.

References

1   Maghnie M, Ghirardello S, Genovese E: Magnetic resonance imaging of the hypothalamus-pituitary unit in children suspected of hypopituitarism: who, how and when to investigate. J Endocrinol Invest 2004;27:496–509.

2   Di Iorgi N, Allegri AE, Napoli F, Bertelli E, Olivieri I, Rossi A, Maghnie M: The use of neuroimaging for assessing disorders of pituitary development. Clin Endocrinol (Oxf) 2012;76:161–176.

3   Fujisawa I, Kikchi K, Nishimura K, Togashi K, Itoh K, Noma S, Minami S, Sagoh T, Hiraoka T, Momoi T: Transection of the pituitary stalk: development of an ectopic posterior lobe associated with MR imaging. Radiology 1987;165:487–489.

4   Kikuchi K, Fujisawa I, Momoi T, Yamanaka C, Kaji M, Nakano Y, Konishi J, Mikawa H, Sudo M: Hypothalamic-pituitary function in growth hormone-deficient patients with pituitary transection. J Clin Endocrinol Metab 1988;67:817–823.

5   Maghnie M, Triulzi F, Larizza D, Scotti G, Beluffi G, Cecchini A, Severi F: Hypothalamic-pituitary dwarfism: comparison between MR imaging and CT findings. Pediatr Radiol 1990;20:229–235.

6   Maghnie M, Triulzi F, Larizza D, Preti P, Priora C, Scotti G, Severi F: Hypothalamic-pituitary dysfunction in growth hormone-deficient patients with pituitary abnormalities. J Clin Endocrinol Metab 1991;72:79–83.

7   Triulzi F, Scotti G, di Natale B, Pellini C, Lukezic M, Scognamiglio M, Chiumello G: Evidence of a congenital midline brain anomaly in pituitary dwarfs: a magnetic resonance imaging study in 101 patients. Pediatrics 1994;93:409–416.

8   Pinto G, Netchine I, Sobrier ML, Brunelle F, Souberbielle JC, Brauner R: Pituitary stalk interruption syndrome: a clinical-biological-genetic assessment of its pathogenesis. J Clin Endocrinol Metab 1997;82:3450–3454.

9   Maghnie M, Genovese E, Villa A, Spagnolo L, Campan R, Severi F: Dynamic MRI in the congenital agenesis of the neural pituitary stalk syndrome: the role of the vascular pituitary stalk in predicting residual anterior pituitary function. Clin Endocrinol (Oxf) 1996;45:281–290.

10 Arrigo T, De Luca F, Maghnie M, Blandino A, Lombardo F, Messina MF, Wasniewska M, Ghizzoni L, Bozzola M: Relationships between neuroradiological and clinical features in apparently idiopathic hypopituitarism. Eur J Endocrinol 1998;139:84–88.

11 Chen S, Leger J, Garel C, Hassan M, Czernichow P: Growth hormone deficiency with ectopic neurohypophysis: anatomical variations and relationship between the visibility of the pituitary stalk asserted by magnetic resonance imaging and anterior pituitary function. J Clin Endocrinol Metab 1999;84:2408–2413.

12 Maghnie M, Ambrosini L, Cappa M, Pozzobon G, Ghizzoni L, Ubertini MG, di Iorgi N, Tinelli C, Pilia S, Chiumello G, Lorini R, Loche S: Adult height in patients with permanent growth hormone deficiency with and without multiple pituitary hormone deficiencies. J Clin Endocrinol Metab 2006;91:2900–2905.

13 Maghnie M, Strigazzi C, Tinelli C, Autelli M, Cisternino M, Loche S, Severi F: Growth hormone (GH) deficiency (GHD) of childhood onset: reassessment of GH status and evaluation of the predictive criteria for permanent GHD in young adults. J Clin Endocrinol Metab 1999;84:1324–1328.

14 Murray PG, Hague C, Fafoula O, Gleeson H, Patel L, Banerjee I, Raabe AL, Hall CM, Wright NB, Amin R, Clayton PE: Likelihood of persistent GH deficiency into late adolescence: relationship to the presence of an ectopic or normally sited posterior pituitary gland. Clin Endocrinol (Oxf) 2009;71:215–219.

15 Cox TD, Elster AD: Normal pituitary gland: changes in shape, size, and signal intensity during the 1st year of life at MR imaging. Radiology 1991;179:721–724.

16 Dietrich RB, Lis LE, Greensite FS, Pitt D: Normal MR appearance of the pituitary gland in the first 2 years of life. Am J Neuroradiol 1995;16:1413–1419.

17 Tsunoda A, Okuda O, Sato K: MR height of the pituitary gland as a function of age and sex: especially physiological hypertrophy in adolescence and in climacterium. Am J Neuroradiol 1997;18:551–554.

18 Argyropoulou M, Perignon F, Brunelle F, Brauner R, Rappaport R: Height of normal pituitary gland as a function of age evaluated by magnetic resonance imaging in children. Pediatr Radiol 1991;21: 247–249.

19 Elster AD: Modern imaging of the pituitary. Radiology 1993;187:1–14.

20 Fink AM, Vidmar S, Kumbla S, Pedreira CC, Kanumakala S, Williams C, Carlin JB, Cameron FJ: Age-related pituitary volumes in prepubertal children with normal endocrine function: volumetric magnetic resonance data. J Clin Endocrinol Metab 2005;90:3274–3278.

21 Marziali S, Gaudiello F, Bozzao A, Scirè G, Ferone E, Colangelo V, Simonetti A, Boscherini B, Floris R, Simonetti G: Evaluation of anterior pituitary gland volume in childhood using three-dimensional MRI. Pediatr Radiol 2004;34:547–551.

22 Takano K, Utsunomiya H, Ono H, Ohfu M, Okazaki M: Normal development of the pituitary gland: assessment with three-dimensional MR volumetry. Am J Neuroradiol 2004;20:312–315.

23 Delman BN: Imaging of pediatric pituitary abnormalities. Endocrinol Metab Clin North Am 2009;38: 673–698.

24 Seeger JF: Normal variations of the skull and its content; in Zimmerman RA, Gibby WA, Carmody RF (eds): Neuroimaging, Clinical and Physical Principles. New York, Springer, 2000, pp 415–489.

25 Bjerre P: The empty sella. A reappraisal of etiology and pathogenesis. Acta Neurol Scand Suppl 1990; 130:1–25.

26 Cacciari E, Zucchini S, Ambrosetto P, Tani G, Carlà G, Cicognani A, Pirazzoli P, Sganga T, Balsamo A, Cassio A: Empty sella in children and adolescents with possible hypothalamic-pituitary disorders. J Clin Endocrinol Metab 1994;78:767–771.

27 Degnan AJ, Levy LM: Pseudotumor cerebri: brief review of clinical syndrome and imaging findings. Am J Neuroradiol 2011;32:1986–1993.

28 de Graaff LC, Baan J, Govaerts LC, Hokken-Koelega AC: Facial and pituitary morphology are related in Dutch patients with GH deficiency. Clin Endocrinol (Oxf) 2008;69:112–116.

29 di Iorgi N, Secco A, Napoli F, Calandra E, Rossi A, Maghnie M: Developmental abnormalities of the posterior pituitary gland. Endocr Dev 2009;14: 83–94.

30 Murray PG, Hague C, Fafoula O, Gleeson H, Patel L, Banerjee I, Raabe AL, Hall CM, Wright NB, Amin R, Clayton PE: Likelihood of persistent GH deficiency into late adolescence: relationship to the presence of an ectopic or normally sited posterior pituitary gland. Clin Endocrinol (Oxf) 2009;71: 215–219.

31 Werder EA, Zachmann M, Wichmann W, Valavanis A: Neurohypophyseal ectopy in growth hormone insufficiency. Horm Res 1989;31:210–212.

32 Genovese E, Maghnie M, Beluffi G, Villa A, Sammarchi L, Severi F, Campani R: Hypothalamic-pituitary vascularization in pituitary stalk transection syndrome: is the pituitary stalk really transected? The role of gadolinium-DTPA with spin-echo T1 imaging and turbo-FLASH technique. Pediatr Radiol 1997;27:48–53.

33 Zenaty D, Garel C, Limoni C, Czernichow P, Léger J: Presence of magnetic resonance imaging abnormalities of the hypothalamic-pituitary axis is a significant determinant of the first 3 years growth response to human growth hormone treatment in prepubertal children with nonacquired growth hormone deficiency. Clin Endocrinol (Oxf) 2003;58: 647–652.

34 Coutant R, Rouleau S, Despert F, Magontier N, Loisel D, Limal JM: Growth and adult height in GH-treated children with nonacquired GH deficiency and idiopathic short stature: the influence of pituitary magnetic resonance imaging findings. J Clin Endocrinol Metab 2001;86:4649–4654.

35 Adan L, Souberbielle JC, Brauner R: Diagnostic markers of permanent idiopathic growth hormone deficiency. J Clin Endocrinol Metab 1994;78: 353–358.

36 Clayton PE, Cuneo RC, Juul A, Monson JP, Shalet SM, Tauber M: Consensus statement on the management of the GH-treated adolescent in the transition to adult care. Eur J Endocrinol 2005;152: 165–170.

37 Ho KK: Consensus guidelines for the diagnosis and treatment of adults with GH deficiency II: a statement of the GH Research Society in association with the European Society for Pediatric Endocrinology, Lawson Wilkins Society, European Society of Endocrinology, Japan Endocrine Society, and Endocrine Society of Australia. Eur J Endocrinol 2007;157:695–700.

38 Leger J, Danner S, Simon D, Garel C, Czernichow P: Do all patients with childhood-onset growth hormone deficiency (GHD) and ectopic neurohypophysis have persistent GHD in adulthood? J Clin Endocrinol Metab 2005;90:650–656.

39 Maghnie M, Aimaretti G, Bellone S, Baldelli R, de Sanctis C, Gargantini L, Gastaldi R, Ghizzoni L, Secco A, Tinelli C, Ghigo E: Diagnosis of GH deficiency in the transition period: accuracy of insulin tolerance test and insulin-like growth factor-I measurement. Eur J Endocrinol 2005;152:589–596.

40 Secco A, di Iorgi N, Napoli F, Calandra E, Calcagno A, Ghezzi M, Frassinetti C, Fratangeli N, Parodi S, Benassai M, Leitner Y, Gastaldi R, Lorini R, Maghnie M, Radetti G: Reassessment of the growth hormone status in young adults with childhood-onset growth hormone deficiency: reappraisal of insulin tolerance testing. J Clin Endocrinol Metab 2009;94:4195–4204.

41 di Iorgi N, Secco A, Napoli F, Tinelli C, Calcagno A, Fratangeli N, Ambrosini L, Rossi A, Lorini R, Maghnie M: Deterioration of growth hormone (GH) response and anterior pituitary function in young adults with childhood-onset GH deficiency and ectopic posterior pituitary: a two-year prospective follow-up study. J Clin Endocrinol Metab 2007; 92:3875–3884.

42 Maghnie M, Uga E, Temporini F, Di Iorgi N, Secco A, Tinelli C, Papalia A, Casini MR, Loche S: Evaluation of adrenal function in patients with growth hormone deficiency and hypothalamic-pituitary disorders: comparison between insulin-induced hypoglycemia, low-dose ACTH, standard ACTH and CRH stimulation tests. Eur J Endocrinol 2005;152:735–741.

43 Secco A, Allegri AE, di Iorgi N, Napoli F, Calcagno A, Bertelli E, Olivieri I, Pala G, Parodi S, Gastaldi R, Rossi A, Maghnie M: Posterior pituitary (PP) evaluation in patients with anterior pituitary defect associated with ectopic PP and septo-optic dysplasia. Eur J Endocrinol 2011;165:411–420.

Mohamad Maghnie, MD, PhD
Department of Pediatrics, IRCCS G. Gaslini, University of Genova
Largo Gerolamo Gaslini, 5
IT–16147 Genova (Italy)
Tel. +39 010 5636574, E-Mail mohamadmaghnie@ospedale-gaslini.ge.it

Mullis P-E (ed): Developmental Biology of GH Secretion, Growth and Treatment.
Endocr Dev. Basel, Karger, 2012, vol 23, pp 30–41 (DOI: 10.1159/000341739)

Spectrum of Insulin-Like Growth Factor Deficiency

Jan M. Wit[a] · W. Oostdijk[a] · M. Losekoot[b]

[a]Department of Pediatrics and [b]Center for Human and Clinical Genetics, Leiden University Medical Center,
Leiden, The Netherlands

Abstract

There are eight known genetic causes of short stature characterized by low serum IGF-1 (IGF-1 deficiency, IGFD) and normal GH secretion. One of these (*GHSR* defect) is a form of secondary IGFD, although the GH peak in provocation tests can be normal. Bioactive GH (*GH1* mutations) can disturb GH secretion, but also GH binding and signaling. The remaining conditions are classified as primary IGFD (GH insensitivity). The clinical phenotype of GH receptor (*GHR*) defects is variable. Of the three GH signal transduction defects, a *STAT5B* defect is well established, but abnormalities in the MAPK pathway (such as *PTPN11* mutations in Noonan syndrome) and NF-κB pathway (IκBα mutation) may also cause IGFD. Homozygous *IGFALS* defects are relatively common, and lead to moderate growth failure, very low serum IGF-1 and even lower IGFBP-3, while a heterozygous *IGFALS* mutation decreases height by 1 SD. Most cases with a homozygous *IGF1* defect are very short, microcephalic, and deaf, but heterozygous mutations may also lead to short stature. IGFD can also have a digenic or oligogenic origin. The diagnostic yield of genetic testing in children with a height <–2.5 SDS and a serum IGF-1 <–2 appears sufficient to perform genetic tests for known candidate genes.

Insulin-like growth factors (IGFs) include two related proteins: IGF-1 and IGF-2. The term IGF deficiency (IGFD) usually refers to a deficiency of IGF-1. There are only few reports on a deficiency of serum IGF-2, for example in Laron syndrome, homozygous *IGFALS* defects, tumor induced cachexia, and liver disorders. In patients with Silver-Russell syndrome demethylation at the IGF2/H19 locus on 11p15 is suggestive of a decreased expression of *IGF2*, and *IGF2* expression is indeed decreased in patients' fibroblasts in culture. However, IGF-2 serum levels were found to be normal [1]. In this review, we shall only consider IGF-1 deficiency.

IGFD can be divided into primary IGFD (a low serum IGF-1 in the presence of normal or elevated growth hormone (GH) secretion) or secondary IGFD (caused by decreased GH secretion, or other conditions associated with a low serum IGF-1 level, such as malnutrition). In this review, we shall not elaborate on classical forms of GH deficiency.

IGF-1 is produced in almost cell types, but serum IGF-1 is mainly derived from the liver. Serum IGF-1 is strongly dependent on age, as well as on pubertal stage: IGF-1 increases with age up to young adulthood, and then slowly decreases again. Serum IGF-1 is also dependent on the concentration of IGF-binding proteins (IGFBPs), particularly IGFBP-3. The majority of circulating IGF-1 molecules are bound to IGFBP-3 and acid-labile subunit (ALS) in a ternary complex. A smaller proportion is bound to IGFBP-3 and other binding proteins as binary complexes, and the remaining small fraction consists of free IGF-1. Thus, a low serum IGFBP-3 level decreases circulating IGF-1 concentration.

Unfortunately, there is a lack of standardization of IGF-1 assays [2]. Ideally, each laboratory should collect its own reference samples and age reference curves, with adequate mathematical techniques to correct for skewness. In the absence of these, the most suitable published reference for the pertinent assay can be used. Assuming that the serum sample is taken from a well-nourished individual, that the assay is well-validated, and that appropriate age references are available, the individual IGF-1 level can be compared with percentiles or SDS (standard deviation score) lines in the appropriate age reference. For group analysis, expression of the result as SDS for age is most useful. Provided that all these conditions are met, IGF-1 deficiency can be defined as a serum IGF-1 level below –2 SDS for age. In adolescence, a further refinement can be made by comparing IGF-1 to specific references for pubertal stage [3].

At present there are 8 reported genetic causes of short stature associated with a low serum IGF-1 and apparently normal GH secretion. A *GHSR* (ghrelin receptor) defect is theoretically a special form of GH insufficiency, but can present as idiopathic short stature (ISS). The bioinactive GH syndrome (Kowarski syndrome), caused by *GH1* mutations, can be associated with diminished GH secretion (GH deficiency type II), but also with GH insensitivity. The remaining 6 conditions, defects of *GHR, STAT5B, PTPN11*, IκBα (*IKGKB*), *IGFALS*, and *IGF1,* are classified as primary IGFD. In the following sections we shall review the recent findings on these eight conditions. A summary of clinical and biochemical features is shown in table 1.

GHSR Defects

GHSR (ghrelin receptor) defects associated with short stature have been reported in five papers (table 2). In the first [4], two missense variants were described, one of which was associated with short stature. A French group [5] reported on two unrelated families (both originating from Morocco) with a *GHSR* variant (c.611C>A, p.A204E). In family 1 the proband (with a homozygous mutation) had idiopathic short stature (ISS), and 4 of the 5 heterozygous relatives were short (table 2). In family 2 the heterozygous proband had isolated GH deficiency (height –3.2 SDS), but not all heterozygous relatives were short. The authors suggested that this mutation is dominant with a penetrance of 66%. In a later publication of this group [6], a boy with

Table 1. Characteristics of causes of primary IGF-1 deficiency

Gene defect/phenotype	GHSR	Bioinactive GH	GHR	STAT5b	IκBα (IKBKB)	PTPN11	IGFALS	IGF-1
Height SDS	–3.7 to +1.1	–7.2 to –0.2	–10 to –2	–10 to –5	–4 to –2	–5 to –2	–4 to 0	–9 to –5
Head circumference	?	normal	normal	normal	normal	normal	–/+	–8 to –5 SDS
Pubertal delay	+	?	+/–	+/–	–	+/–	+	–/+
Midface hypoplasia	–	–	+/–	+/–	–	–	–	–
Deafness	–	–	–	–	–	–	–	+/–
Intellectual delay	–	–	–	–	–	–/+	–	+
Immune deficiency	–	–	–	+	+	–	–	–
Hypoglycemia	–	–	+	–/+	–	–	–	–
Hyperinsulinemia	?	–	–	–	–	–	+	+/–
IGF-1	low or normal	–3.4 to –1 SDS	low	low	–3.5 to –1.5 SDS	low or normal	–11 to –3 SDS	low, normal or high
IGFBP-3	low or normal	–2.7 to –0.5 SDS	low	low	–5 to –2 SDS	low or normal	–19 to –3 SDS	normal
ALS	?	?	low	low	?	low or normal	undetectable	normal
GH_{max}, ng/ml	low or normal	5–261	high	high/normal	high	normal	high or normal	high or normal
GHBP deficiency	?	–	+/–	–	?	–	–	–
Prolactin	?	normal	normal or slightly increased	High	normal	normal	normal	normal
Homozygous or compound heterozygous	+	+	+	+	?	–	+	+
Heterozygous	+/–	+	–	–/+	+	+	–/+	–/+

+ = Present; – = absent; +/– = mostly present, sometimes absent; –/+ = mostly absent, sometimes present; ? = unknown.

Wit · Oostdijk · Losekoot

partial isolated GH deficiency was found to be compound heterozygous for two variants (table 2). The fourth paper [7] described four novel heterozygous mutations in 127 short children with GH deficiency or ISS (table 2). Most recently, two functionally relevant *GHSR* variants in 31 children with constitutional delay of growth and puberty were found [8], showing a decrease in basal receptor activity in part explained by a reduction in cell surface expression. Both patients were short in adolescence but of normal adult stature, had a normal GH provocation test, and serum IGF-1 levels in the lower half of the normal range or slightly below (table 2). It is noteworthy that two siblings of the first patient were also heterozygous for the variant, but of normal stature. The father and sister of the second patient (also heterozygous carriers of the mutation) had delayed puberty with a normal adult height.

Taking all these findings together, we conclude that it is still uncertain if *GHSR* variants cause short stature, and what the phenotype is of *GHSR* mutations. The functional relevance of several of the described variants in vivo is uncertain, not all carriers share the same phenotype, and GH secretion and serum IGF-1 were not low in all cases (table 2).

Bioinactive Growth Hormone (Kowarski Syndrome)

Only few cases with a putative bioinactive GH, due to a *GH1* mutation, have been reported [9]. The clinical, biochemical and genetic features are summarized in table 3. The reader should realize that according to HGVS nomenclature the location of the reported mutations should be moved upwards by 26 codons. Most cases with this syndrome show a normal or slightly increased GH secretion, but some cases present with type II GH deficiency in combination with decreased GH bioactivity (table 3). Serum IGF-1 is usually low and catch-up growth on GH replacement therapy is moderate or good [9].

As previously reviewed [9, 10], a heterozygous missense mutation (p.R77C, or R103C) was found by a Japanese group as well as by a Swiss group. Although this variant was also present in the normal-statured father, the reduced capability to induce GHR/GHBP gene transcription suggests that this variant may have functional consequences [9]. Another heterozygous missense mutation (p.D112G, or D138G) that prevents dimerization of the GHR was detected by the Japanese group. For the homozygous missense mutation (p.C53S, or C79S) leading to disruption of one of the disulfide bonds, it was shown that GHR binding and JAK2/STAT5 signaling activity was reduced (for review, see [9]). In a child with severe short stature caused by the autosomal-dominant from of GH deficiency (IGHD-2), the mutant GH (R178H, or R204H) was not only shown to decrease GH secretion but also its action [11]. Finally, a heterozygous missense mutation (p.P59L, or p.P85L) was shown to cause partial GH deficiency combined with bioinactive GH syndrome [12]. We recently detected another mutation at the same location (p.P59S, or p.P85S) in two siblings with bioinactive GH syndrome [Petkovic et al., in prep.].

Table 2. Clinical and biochemical characteristics in patients with *GHSR* mutations

Reference	Mutation (protein)*	Homozygous/ heterozygous	Patient No.	Height SDS
Wang et al. [4], 2004	F279L	heterozygous	1	<–2.0
Pantel et al. [5], 2006	A204E	homozygous	A.II.2	–3.7
		heterozygous	A.II.1	–1.1
		heterozygous	A.II.3	–2.0
		heterozygous	A.II.4	–2.2
		heterozygous	A.I.1	–2.7
		heterozygous	A.I.2	–3.7
		heterozygous	B.II.1	–3.2
		heterozygous	B.I.2	–2.0
		heterozygous	B.II.2	–1.2
		heterozygous	B.II.3	+1.0
Pantel et al. [6], 2009	W2X R237W	compound heterozygous	II.1	–2.7
Inoue et al. [7], 2011	ΔQ36**	heterozygous	AII.1	–3.1
			A2.II.2	–3.4
			A3.II.3	–3.4
	P108L	heterozygous	BII.2	–2.8
	C173R	heterozygous	CII.1	–2.8
		heterozygous	CI.2	–1.9
	D246A	heterozygous	DII.1	–3.3
		heterozygous	DII.2	–3.1
		heterozygous	DI.2	–3.0
Pugliese-Pires et al. [8], 2011	S84I	heterozygous	1	–2.4→–0.7
	V182A	heterozygous	2	–2.3→–1.4

ISS = Idiopathic short stature; IGHD = isolated growth hormone deficiency; CDGP = constitutional delay of growth and puberty.
*NM_198407.2 is used as reference sequence. The amino acid numbering is identical for HGVS nomenclature.
** ΔQ36 is c.107_109del, p.Gln36del according to HGVS nomenclature.

Growth Hormone Receptor Defects

In contrast to the previous two conditions, there is little doubt that genetic defects of the *GHR* can cause a severe postnatal growth failure, with a height SDS down to –10. The classical form of this condition (Laron syndrome, growth hormone Insensitivity Syndrome), with very low serum IGF-1 despite increased GH secretion, was first described by Laron, but thereafter approximately 300 cases have been identified.

GH_{max}, ng/ml	IGF-1 (SDS)	Diagnosis	Prevalence	GH therapy, response
		ISS	1/43	
29; 48	normal	ISS	1/41 ISS, 1/51 IGHD	+, good
				−
14; 9	normal	ISS		+, good
43	normal	ISS		−
		ISS		−
		ISS		−
3; 5	low	IGHD		+, good
		ISS?		−
				−
				−
6; 7	−2.1	IGHD	?	+, good
		ISS or IGHD	4/113 ISS+14 IGHD	?
		ISS or IGHD		−
		ISS or IGHD		−
		ISS or IGHD		−
		ISS or IGHD		−
				−
		ISS or IGHD		−
		ISS or IGHD		−
		ISS or IGHD		−
10.3	−2.3; −0.5	CDGP	2/31 CDGP	−
7.9	−1.3	CDGP		−

For up to date information about this syndrome, the reader is referred to two recent reviews [9, 10]. In the majority of cases with mutations and in the rare cases with a deletion, GHBP (the extracellular part of the GHR) is not formed, so that serum GHBP deficiency can be used as a diagnostic marker. However, mutations affecting the transmembrane or intracellular part of the GHR are associated with normal or even elevated serum GHBP levels. For clinical practice, it is important to note that even moderate short stature (up to −2 SDS) can be caused by GHR variants [10].

Table 3. Clinical and biochemical characteristics in patients with presumed bioinactive growth hormone syndrome (*GH1* mutations)

Reference	Mutation (p)*	Homozygous/ heterozygous	Patient no.	Height SDS	GH$_{max}$, ng/ml	IGF-1 SDS	IGFBP-3 SDS	GH therapy/ response
Takahashi, 1996	R77C	heterozygous	II.2	–6.1	38; 15; 35	–2.1	–1.0?	+/poor
	R77C	heterozygous	I.1	–0.2	23.7	–1	–1.4?	–
Takahashi, 1997	D112G	heterozygous	II.2	–3.6	26; 41; 51	–2.5		+/good
Millar, 2003	T27I	heterozygous		normal				?
	K41R	heterozygous		?				?
	N47D	heterozygous		?				?
	S71F	heterozygous		?				?
	S108R	heterozygous		?				?
	T175A	heterozygous		?				?
Besson, 2005	C53S	homozygous	II.2	–3.6	44.7	–3.4	–0.7	+/good
Petkovic, 2007	R77C	heterozygous	III.3	–2.5→–1.9	253	–3.1 to –2.2	–1.0 to –0.5	–
		heterozygous	II.3	–1.4	261	–2.1	–1.9	–
		heterozygous	I.1	–1.7	242	–2.3	–1.7	–
Petkovic, 2010	R178H	heterozygous		–6.0→–7.2	3.9	–2.3	–2.7	+/moderate (+3.2 SD)
Petkovic, 2011	P59L	heterozygous	II.1	–2.5	6.9; 5.6	–2.4		+/good
		heterozygous	II.2	–2.0	4.7; 4.8			+/good
		heterozygous	I.7	–4.2				–

* Mutations are indicated as published. For the HGVS nomenclature, the NM_000515.3 reference sequence is used and 26 amino acids have to be added to the numbers in the table (e.g. R77C: HGVS c.307C>T p.R103C).

STAT5B Mutations

The first discovered example of a disorder in GH signal transduction was a homozygous *STAT5B* mutation. In the last 9 years 7 homozygous mutations in 10 cases have been described (for reviews, see [13] and [10]). The growth curves of these patients are very similar to those of severe *GHR* defects (height SDS between –5 and –10), but in all except one case there is another specific clinical feature: a severe immune dysfunction, characterized by hemorrhagic varicella, chronic pulmonary disease, lung fibrosis, lymphocytic interstitial pneumonia, and severe eczema. One case only showed a very mild immunophenotype (limited to hemorrhagic varicella) [14], which is probably caused by some remaining functionality of a truncated STAT5B protein [Nadeau, pers. commun.]. An elevated serum prolactin was found in all cases.

Until recently, it was uncertain if also a heterozygous *STAT5B* mutation could cause short stature, because not all obligate heterozygous parents were short. However, we recently reported on two siblings with a heterozygous *STAT5B* mutation who had the classical picture of GH insensitivity, and another short child with heterozygous *STAT5B* mutation in combination with an *IGFALS* mutation with a similar clinical picture [15].

Activating *PTPN11* Mutations

Given the complexity of the GH signaling pathway, containing at least three distinct (but interrelated) routes, one can confidently postulate that variants in other genes can also be related to GH insensitivity. In fact, it was shown that activation of the RAS-MAPK pathway by mutations of SHP-2 (encoded by *PTPN11*) or of other components of this pathway, as seen in Noonan syndrome and related syndromes, leads to dephosphorylation of STAT5B, causing downregulation of its activity and a partial GH insensitivity [10, 16]. However, the partial GH insensitivity cannot fully explain the decreased body stature of these patients.

IκBα Mutation

We have recently provided evidence, from observations in two children with severe short stature and other abnormalities, that also the NF-κB pathway plays a role in GH signalling. The first case had a heterozygous mutation in *IKBKB* (MIM 603258), which causes a severe immunodeficiency. Endocrine workup because of severe short stature showed a severe GH insensitivity [17]. The second case had a 17q21–25 duplication associated with GH insensitivity and disturbed STAT5B, PI3K and NF-κB signaling, possibly due to *PRKCA* (MIM 176960) mRNA overexpression [18].

ALS Deficiency

Another cause of IGFD is a decreased serum ALS. The first case of ALS deficiency caused by a homozygous mutation of *IGFALS* was published in 2004, and this was followed by many other reports (reviewed in [19]). The effect of a complete *IGFALS* defect on growth is approximately 2–3 SD, but the first reported case showed a growth pattern consistent with delayed growth and puberty, resulting in an adult height within the normal range. Clinical and biochemical features of individuals with homozygous mutations include moderate short stature, usually between –2 and

−4 SDS, delayed puberty (in approximately 50% of the affected males), decreased serum IGF-1 and even more decreased serum IGFBP-3. Heterozygous mutations of *IGFALS* also have some negative effect on height, but usually only 1 SD in comparison to wild-type controls [20]. In such cases serum ALS is usually not far below the reference range, in contrast to undetectable serum levels in patients with homozygous mutations.

In some cases we have observed decreased serum ALS levels and reduced 150kD ternary complex formation as assessed by column chromatography, not associated with pathogenic *IGFALS* mutations in the coding region [15]. This suggests that decreased ALS production can be caused by abnormalities outside the coding region of *IGFALS*, for example in the promoter region.

IGF Defects

The first published case with an *IGF1* defect had a partial gene deletion (exons 4 and 5). He was mentally retarded and deaf, and had prenatal and postnatal growth failure, microcephaly, and no detectable serum IGF-1, but a normal serum IGFBP-3 [21]. It took 9 years before the next confirmed case was published. This patient had a homozygous missense mutation with strongly decreased binding affinity to the IGF-1 receptor (IGF1R) [22] (thus a form of bioinactive IGF-1), and a similar phenotype as the first patient. In this case, serum (mutant) IGF-1 was strongly elevated. The third confirmed case had a milder phenotype (no deafness, less short) caused by a missense mutation with less effect on receptor binding [23]. There may be a fourth case, but the putative causative mutation was later found to be a polymorphism. More details on *IGF1* defects can be found in a recent review [24].

In our report on the second case, we provided evidence that heterozygosity for an *IGF1* defect has a mild effect on growth and head circumference [22]. More recently, we showed that a heterozygous mutation of *IGF1* can also cause a more severe growth failure, particularly if combined with a short genetic background [25].

Digenic and Oligogenic Causality

So far, the scientific community has been focused on discovering monogenetic causes of IGF-1 deficiency, and this approach has been quite successful. However, we have recently provided evidence that IGF-1 deficiency can also be caused by abnormalities in two genes (digenic causality) [15] and it is plausible that in the future cases will be found with several gene mutations (oligogenic causality). The first case with a putative digenic cause of IGF-1 deficiency had a combination of a heterozygous *STAT5B* mutation and a heterozygous *IGFALS* mutation, and he was

1 SD shorter than his brother who only had the *STAT5B* mutation. The second case also had a combination of a heterozygous *STAT5B* mutation and an *IGFALS* mutation [15].

Diagnostic Approach of Children with IGF-1 Deficiency

In a previous paper we proposed a clinical algorithm for genetic testing in children with a low IGF-1, either born with a normal or low birth size [26], and a similar algorithm was presented by others [10]. Birth size, head circumference, serum IGF-1, IGFBP-3, ALS, prolactin and GHBP, as well as the presence or absence of additional clinical features (mental retardation, deafness, immune disorders) are relevant clinical clues to determine the candidate gene to be tested (table 1). We have suggested that an IGF-1 generation test can be useful to distinguish GH-sensitive forms from GH insensitivity [26]. For detailed information on the clinical, biochemical and genetic findings in patients with various forms of IGFD a website is available (http://www.growthgenetics.com).

However, the apparent rarity of the reported gene defects might withhold clinicians from performing genetic tests in their patients. We have recently investigated the diagnostic yield in a group of 36 patients with varying degrees of short stature and IGF-1 deficiency [15], and shown that if the proposed algorithm is followed (genetic testing if height <–2.5 SDS and serum IGF-1 <–2 SDS), a causative genetic abnormality is found in approximately 30%. In children with less severe short stature and/or modestly decreased serum IGF-1 the likelihood of finding variants in the known genes is much lower. SNP-arrays for the detection of copy number variants have led to novel insights in the possible role of other signal transduction pathways in IGF-1 generation [18], and may elucidate other genes involved. However, whole exome sequencing is expected to be an even better tool for discovering novel genes involved in IGF-1 deficiency.

Conclusion

Genetic testing of children with primary IGF-1 deficiency can lead to the detection of one of the eight known genetic causes, and to the discovery of novel causes. The rapidly expanding use of next gene sequencing will probably lead to the discovery of more monogenic causes of IGF-1 deficiency, as well as digenic and oligogenic causes.

References

1 Binder G, Seidel AK, Martin DD, Schweizer R, Schwarze CP, Wollmann HA, Eggermann T, Ranke MB: The endocrine phenotype in Silver-Russell syndrome is defined by the underlying epigenetic alteration. J Clin Endocrinol Metab 2008;93: 1402–1407.

2 Clemmons DR: Consensus statement on the standardization and evaluation of growth hormone and insulin-like growth factor assays. Clin Chem 2011; 57:555–559.

3 Lofqvist C, Andersson E, Gelander L, Rosberg S, Blum WF, Albertsson WK: Reference values for IGF-1 throughout childhood and adolescence: a model that accounts simultaneously for the effect of gender, age, and puberty. J Clin Endocrinol Metab 2001;86:5870–5876.

4 Wang HJ, Geller F, Dempfle A, Schauble N, Friedel S, Lichtner P, Fontenla-Horro F, Wudy S, Hagemann S, Gortner L, Huse K, Remschmidt H, Bettecken T, Meitinger T, Schafer H, Hebebrand J, Hinney A: Ghrelin receptor gene: identification of several sequence variants in extremely obese children and adolescents, healthy normal-weight and underweight students, and children with short normal stature. J Clin Endocrinol Metab 2004;89:157–162.

5 Pantel J, Legendre M, Cabrol S, Hilal L, Hajaji Y, Morisset S, Nivot S, Vie-Luton MP, Grouselle D, de Kerdanet M, Kadiri A, Epelbaum J, Le Bouc Y, Amselem S: Loss of constitutive activity of the growth hormone secretagogue receptor in familial short stature. J Clin Invest 2006;116:760–768.

6 Pantel J, Legendre M, Nivot S, Morisset S, Vie-Luton MP, Le Bouc Y, Epelbaum J, Amselem S: Recessive isolated growth hormone deficiency and mutations in the ghrelin receptor. J Clin Endocrinol Metab 2009;94:4334–4341.

7 Inoue H, Kangawa N, Kinouchi A, Sakamoto Y, Kimura C, Horikawa R, Shigematsu Y, Itakura M, Ogata T, Fujieda K: Identification and functional analysis of novel human growth hormone secretagogue receptor (GHSR) gene mutations in Japanese subjects with short stature. J Clin Endocrinol Metab 2011;96:E373–E378.

8 Pugliese-Pires PN, Fortin JP, Arthur T, Latronico AC, Mendonca BB, Villares SM, Arnhold IJ, Kopin AS, Jorge AA: Novel inactivating mutations in the GH secretagogue receptor gene in patients with constitutional delay of growth and puberty. Eur J Endocrinol 2011;165:233–241.

9 Mullis PE: Genetics of GHRH, GHRH-receptor, GH and GH-receptor: its impact on pharmacogenetics. Best Pract Res Clin Endocrinol Metab 2011;25:25–41.

10 David A, Hwa V, Metherell LA, Netchine I, Camacho-Hubner C, Clark AJ, Rosenfeld RG, Savage MO: Evidence for a continuum of genetic, phenotypic, and biochemical abnormalities in children with growth hormone insensitivity. Endocr Rev 2011;32:472–497.

11 Petkovic V, Godi M, Pandey AV, Lochmatter D, Buchanan CR, Dattani MT, Eble A, Fluck CE, Mullis PE: Growth hormone (GH) deficiency type II: a novel GH-1 gene mutation (GH-R178H) affecting secretion and action. J Clin Endocrinol Metab 2010;95:731–739.

12 Petkovic V, Eble A, Pandey AV, Betta M, Mella P, Fluck CE, Buzi F, Mullis PE: A novel GH-1 gene mutation (GH-P59L) causes partial GH deficiency type II combined with bioinactive GH syndrome. Growth Horm IGF Res 2011;21:160–166.

13 Hwa V, Nadeau K, Wit JM, Rosenfeld RG: STAT5B deficiency: Lessons from STAT5B gene mutations. Best Pract Res Clin Endocrinol Metab 2011;25: 61–75.

14 Walenkamp MJ, Vidarsdottir S, Pereira AM, Karperien M, van Doorn J, van Duyvenvoorde HA, Breuning MH, Roelfsema F, Kruithof MF, van Dissel J, Janssen R, Wit JM, Romijn JA: Growth hormone secretion and immunological function of a male patient with a homozygous STAT5B mutation. Eur J Endocrinol 2007;156:155–165.

15 Wit JM, van Duyvenvoorde HA, Scheltinga SA, de Bruin S, Hafkenscheid L, Kant SG, Ruivenkamp CAL, Gijsbers ACJ, van Doorn J, Feigerlova E, Noordam C, Walenkamp MJ, Claahsen-van de Grinten H, Stouthart P, Bonapart IE, Pereira AM, Gosen J, Delemarre-Van de Waal HA, Hwa V, Breuning MH, Domene HM, Oostdijk W, Losekoot M: Genetic analysis of short children with apparent growth hormone insensitivity. Horm Res Paediatr 2012;77:320–333.

16 Binder G: Noonan syndrome, the Ras-MAPK signalling pathway and short stature. Horm Res 2009; 71(suppl 2):64–70.

17 Wu S, Walenkamp MJ, Lankester A, Bidlingmaier M, Wit JM, De Luca F: Growth hormone and insulin-like growth factor I insensitivity of fibroblasts isolated from a patient with an I{kappa} B{alpha} mutation. J Clin Endocrinol Metab 2010; 95:1220–1228.

18 Mul D, Wu S, de Paus RA, Oostdijk W, Lankester AC, Duyvenvoorde HA, Ruivenkamp CA, Losekoot M, Tol MJ, De LF, Van D, Wit JM: A mosaic de novo duplication of 17q21–25 is associated with GH insensitivity, disturbed in vitro CD28-mediated signaling, and decreased STAT5B, PI3K, and NF-kappaB activation. Eur J Endocrinol 2012;166: 743–752.

19 Domene HM, Hwa V, Jasper HG, Rosenfeld RG: Acid-labile subunit (ALS) deficiency. Best Pract Res Clin Endocrinol Metab 2011;25:101–113.

20 Fofanova-Gambetti OV, Hwa V, Wit JM, Domene HM, Argente J, Bang P, Hogler W, Kirsch S, Pihoker C, Chiu HK, Cohen L, Jacobsen C, Jasper HG, Haeusler G, Campos-Barros A, Gallego-Gomez E, Gracia-Bouthelier R, van Duyvenvoorde HA, Pozo J, Rosenfeld RG: Impact of heterozygosity for acid-labile subunit (IGFALS) gene mutations on stature: results from the international acid-labile subunit consortium. J Clin Endocrinol Metab 2010;95: 4184–4191.

21 Woods KA, Fraser NC, Postel Vinay MC, Savage MO, Clark AJ: A homozygous splice site mutation affecting the intracellular domain of the growth hormone (GH) receptor resulting in Laron syndrome with elevated GH-binding protein [see comments]. J Clin Endocrinol Metab 1996;81: 1686–1690.

22 Walenkamp MJ, Karperien M, Pereira AM, Hilhorst-Hofstee Y, van Doorn J, Chen JW, Mohan S, Denley A, Forbes B, van Duyvenvoorde HA, van Thiel SW, Sluimers CA, Bax JJ, de Laat JA, Breuning MB, Romijn JA, Wit JM: Homozygous and heterozygous expression of a novel insulin-like growth factor-I mutation. J Clin Endocrinol Metab 2005;90:2855–2864.

23 Netchine I, Azzi S, Houang M, Seurin D, Perin L, Ricort JM, Daubas C, Legay C, Mester J, Herich R, Godeau F, Le Bouc Y: Partial primary deficiency of insulin-like growth factor (IGF)-I activity associated with IGF1 mutation demonstrates its critical role in growth and brain development. J Clin Endocrinol Metab 2009;94:3913–3921.

24 Netchine I, Azzi S, Le BY, Savage MO: IGF1 molecular anomalies demonstrate its critical role in fetal, postnatal growth and brain development. Best Pract Res Clin Endocrinol Metab 2011;25:181–190.

25 van Duyvenvoorde HA, van Setten PA, Walenkamp MJ, van DJ, Koenig J, Gauguin L, Oostdijk W, Ruivenkamp CA, Losekoot M, Wade JD, De MP, Karperien M, Noordam C, Wit JM: Short stature associated with a novel heterozygous mutation in the insulin-like growth factor 1 gene. J Clin Endocrinol Metab 2010;95:E363–E367.

26 Wit JM, Kiess W, Mullis P: Genetic evaluation of short stature. Best Pract Res Clin Endocrinol Metab 2011;25:1–17.

Prof. Jan M. Wit
Department of Pediatrics, J6S, Leiden University Medical Center
PO Box 9600
NL–2300 RC Leiden (The Netherlands)
Tel. +31 71 5262824, E-Mail j.m.wit@lumc.nl

Mullis P-E (ed): Developmental Biology of GH Secretion, Growth and Treatment.
Endocr Dev. Basel, Karger, 2012, vol 23, pp 42–51 (DOI: 10.1159/000341745)

Downstream Insulin-Like Growth Factor

Roland Pfäffle · Wieland Kiess · Jürgen Klammt

Department of Pediatrics, University of Leipzig Medical School, Leipzig, Germany

Abstract

Until 10 years ago genetic defects that cause a child to be born small for gestational age (SGA) were poorly defined. With the first descriptions of patients born small for gestational age carrying mutations within the insulin-like growth factor type 1 receptor (IGF-1R) gene, genetic defects at the lower end of the GH-IGF-1 axis were identified as a monogenetic cause of intrauterine growth retardation. These patients failed to thrive despite normal IGF-1 serum concentrations thereby establishing a concept of IGF-1 resistance in these patients. The identification of additional patients along with functional, genetic and structural examinations of the different IGF-1R mutations have provided evidence for a variability of the pathogenic impact that mutations of the IGF-1R have on human longitudinal growth. However, the seemingly variable incidence of further clinical features such as developmental delay, suggest that at least some of the functions within the IGF-IGF-1R system are in part redundant. This redundancy may depend on the genetic background and environmental factors. At the lower end of the GHRH-IGF-1 axis, primary IGF-1 deficiency and IGF-1 resistance due to defects within the IGF-1 and IGF-1 receptor (IGF-1R) genes account for approximately 10–15% of all cases with intrauterine and postnatal growth retardation.

Growth is the result of a close interplay of humoral, cellular and extracellular factors. In the somatotropic axis, the insulin-like growth factor (IGF) 1 signal is transmitted to the interior of the cell by the type 1 IGF receptor (IGF-1R).

The existence of a specific receptor for IGF-1 (somatomedin C) was already shown in the mid-1970s prompting Nissley and Rechler [1] in 1984 to the prediction that 'the demonstration of Sm/IGF receptors. . . will make it possible to investigate patients who are candidates for having end-organ resistance to Sm/IGF'. However, although the IGF-1R cDNA was published in 1986 [2], it took almost another 20 years until the first description of 2 patients carrying IGF-1R mutations as the cause of growth retardation due to IGF-1 insensitivity [3].

IGF-1 has the central role in the regulation of growth. It mediates most of the growth-promoting effects of growth hormone (GH). Although one could consider GH deficiency (GHD) as a form of secondary IGF-1 deficiency (IGFD), there are substantial differences in the clinical phenotypes of secondary IGFD resulting from either GHD or GH resistance as well as phenotypes of primary IGFD due to IGF-1 gene defects or IGF-1 resistance resulting from defects within the IGF-1 receptor (IGF-1R).

The most evident difference in the clinical phenotype is most probably that both primary IGFD and IGF resistance are associated with intrauterine growth retardation and that affected patients are born small for gestational age (SGA).

Gene knockout experiments already suggested such a phenotype more than 10 years ago: although mice with homozygous deletions of the Igf1 gene were about 40% smaller than wild-type mice, some of them survived, whereas mice homozygous for Igf1R knockout were 60% smaller at birth compared to wild-type mice and they invariably died of pulmonary failure [4].

IGF-1 Mutations

Clinically, there have been very impressing descriptions of patients with genetically caused IGF-1 deficiencies. Overall, these cases seem to be rare, but they share a common clinical phenotype that is characterized by intrauterine growth retardation (birth length –2.4 to –6.5 SDS), postnatal growth failure (height –4.9 to –8.0 SDS), psychomotor retardation, sensoneural deafness, micrognathia and delayed puberty [5, 6]. Their endocrine parameters, however, are quite notable too: whereas IGF-1 serum levels are most likely undetectable when there is homozygosity for a complete IGF-1 gene deletion as described by Woods et al. [7], IGF-1 serum levels may be found either elevated, normal or decreased depending on the antibodies used in the immunoassay in patients with bioinactivating amino acid substitutions within the IGF-1 molecule. GH levels seem to be elevated during childhood whereas they are normal in adulthood. IGFBP-3 and IGFALS levels are normal, which is unusual in respect to the very low IGF-1 serum levels. More recently, homozygous mutations of the *IGF1* gene have been described, which show a less severe phenotype [8]. Therefore, genetic defects leading to a partial loss of IGF-1 activity may be more frequent in SGA patients than initially thought. The identification of patients with IGF-1 gene defects holds therapeutic implications, as recombinant IGF-1 would possibly be a therapeutic option.

Molecular Genetics of the IGF-1R

The IGF-1R (OMIM *147370) gene maps to chromosome band 15q26.3 spanning approximately 315 kb. 21 exons encode for an mRNA of 11,242 bp including unusually long stretches of 5′- and 3′-untranslated regions (UTR). The open reading frame of the IGF-1R mRNA comprises 4,104 nucleotides encoding for a protein of 1,367 amino acids. Various RNA-binding proteins and micro-RNA miR-145 bind to IGF-1R 5′- and 3′-UTR thereby differentially regulating mRNA stability and initiation of translation [9]. Although the IGF-1R is basically expressed in almost all cells and tissues, its expression is highly regulated depending on the developmental state, nutritional status and levels of the extracellular hormones. In human embryos, IGF-1R expression can be detected from the 8-cell stage onwards and persists throughout embryonic, fetal and postnatal development thereby declining steadily until adulthood.

During synthesis of the 1,367 amino acids (aa) single-chain preproreceptor and import into the endoplasmic reticulum, the 30-aa signal peptide is removed. The proreceptor monomers become folded, dimerized by disulfide-bonds, and – after core-N-glycosylation (at 16 potential glycosylation sites) – is transported to the trans-Golgi network. Within the Golgi apparatus, the maturation of the proreceptor is completed by terminal glycosylation and cleavage of the proreceptor resulting in the mature $\alpha_2\beta_2$ heterotetrameric glycoprotein [10].

The IGF-1R is activated with highest affinity by its ligand IGF-1, but can be activated also by IGF-2 and supraphysiological doses of insulin [11].

Human IGF-1 Receptor Mutations

The first mutations within the IGF-1R were described in 2003 in a combined paper by researchers from the USA, France and Germany [3]. Since this initial description of 2 patients born small for gestational age (SGA) that presented with IGF-1 resistance and failure to catch-up growth during postnatal development, additional reports have proven the pathogenic impact of heterozygous IGF-1R mutations primarily on human pre- and postnatal growth [12–21].

With the exception of patient 1 reported by Abuzzahab et al. [3] and a recently reported boy from Australia [22] who each bear two compound heterozygous mutations, all the index cases are heterozygous carriers of their respective mutations. Mutations are dispersed throughout the coding sequence. 12 of 14 mutations are single nucleotide substitutions, which result in amino acid substitutions (ten missense) or the introduction of a premature termination codon (two nonsense). In 1 patient, 19 nucleotides within exon 18 were duplicated [14] and in another proband a single basepair was deleted [12] (fig. 1).

For each mutation described so far, evidence of a pathogenic relevance can be derived by comparative, genetic, structural and biochemical studies. In many pedigrees the mutations co-segregated well with the growth-retarded phenotype in the families. This indicates that a familial history of short stature and/or SGA is a valuable indicator for possible genetic defects of the IGF-1R and it means that IGF-1

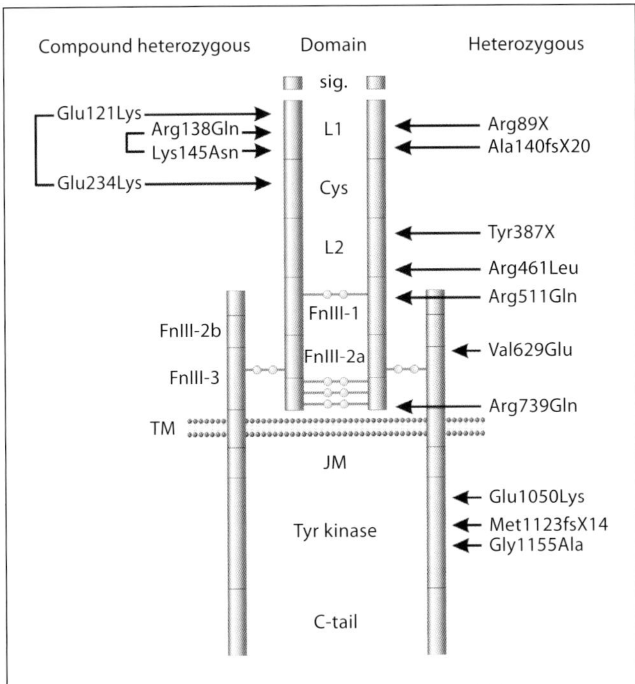

Fig. 1. Heterozygous IGF-1R mutations reported so far are distributed throughout the entire receptor mapping to both α- and β-subunits. Intracellular variants with proven pathogenic relevance have only been described for the tyrosine kinase domain. Two probands were identified with both alleles affected (compound heterozygous). Domain composition according to Adams et al. [10] and UniProtKB (acc. No. P08069). Filled circles = Interchain disulfide bonds.

resistance due to mutations of the IGF-1R most often is an autosomal-dominant-inherited condition.

Mutations Resulting in Haploinsufficiency of the IGF-1R

In 4 patients, IGF-1 resistance can be attributed to a decreased dose of the IGF-1R protein [3, 12, 14, 19]. In 1 of the patients reported by Abuzzahab et al. [3], haploinsufficiency results from inactivation of one allele by a nonsense mutation about 60 amino acids after the signal peptide Arg89X. It seems unlikely that there is any functional interference of the remnant protein with the wild-type receptors or circulating IGFs, because the heavily truncated protein lacks any functional domain. However, the authors showed that transcripts of both the mutant and the wild-type allele are equally expressed in fibroblasts of the affected boy and his mother. On the contrary, the 19 aa duplication identified by Fang et al. [14] results in the introduction of an early stop codon into the mRNA which is then subjected to nonsense-mediated mRNA decay (NMD), which prevents the translation of truncated, potentially deleterious proteins.

Mutations Affecting IGF-1R Biosynthesis

Two mutations of the IGF-1R, Val629Glu [21] and Arg739Gln [17], have been shown to interfere with distinct steps of IGF-1R biosynthesis. Both immunofluorescence staining of the receptor and its precursors and complementary biochemical approaches in transfected cells revealed that substitution of glutamic acid for valine at residue 629 leads to retention of the nascent protein within the endoplasmic reticulum. One can assume that the density of functional receptors at the cell surface of native patient cells is thereby diminished, thus resembling IGF-1R haploinsufficiency. Patient fibroblasts carrying the heterozygous Arg739Gln substitution showed a reduction of IGF-1R on the cell surface by half with similar binding affinities for IGF-1, thereby confirming surface haploinsufficiency for this mutant protein. The Arg739Glu mutation leads to a substantial increase in uncleaved prereceptors at the expense of processed β-subunits.

Similar perturbations resulting in intracellular retention of misfolded IGF-1R peptides have been demonstrated for the recently identified compound heterozygous Glu121Lys and Glu234Lys mutations, although the precise mechanisms remain to be elucidated.

Mutations Interfering with Ligand Binding and Transmembrane Signaling

Patient 1 described by Abuzzahab et al. [3] was the first patient identified who bears two compound heterozygous mutations. Both the Arg138Gln mutation inherited from the father and the Lys145Asn mutation transmitted from the mother affect the L1 (leucine-rich 1) domain of the extracellular portion of the receptor. Arginine 138 has been suggested to be part of the contact interface with IGF-1 [23]. Therefore, this substitution might interfere with IGF-1 binding. The structural impact of the Lys145Asn mutation, however, is less clear.

The Arg511Gln mutation identified in a girl with severe growth retardation was suggested to impair receptor dimerization and conformational reorganization of the receptor after ligand binding [15]. However, arginine 511 is only poorly conserved during evolution which might suggest that this sequence variation might rather predispose to short stature than be a monogenic cause for pathologic growth impairment.

Impaired receptor internalization has been demonstrated for the Arg461Leu mutation located within the L2 domain of the IGF-1R [16]. Interestingly, this mutation does not interfere with IGF-1 ligand binding, but reduces IGF-dependent tyrosine phosphorylation of the receptor and IRS signaling molecules as well as decreases DNA synthesis.

Mutations Disrupting IGF-1R Tyrosine Kinase Activity

Two point mutations within the cytoplasmic receptor portion have been described in 2 female patients with short stature. Although the mutations affect two residues that are 104 aa apart (Glu1050Lys [20] and Gly1155Ala [18]) and reside within distinct structural sub-domains of the tyrosine kinase domain, both amino acids are involved in the coordination of a Mg^{2+} ion within the activated receptor [24]. The quantification of IGF-1-induced receptor autophosphorylation employing patient's fibroblasts (Glu1050Lys) as well as Igf1r-deficient mouse fibroblasts co-expressing wild-type and mutant receptors (Gly1155Ala) interestingly showed a partial dominant effect of the mutant proteins. If one assumes an equal expression of wild-type and mutant alleles and a random distribution of monomeric proreceptors, one would expect only 1/4 of all receptors to be fully functional $(\alpha\beta)_{WT}/(\alpha\beta)_{WT}$ dimers, whereas 3/4 of the assembled receptors would contain at least one mutated hemi-receptor. Such a distribution would explain the observed phenotype. However, detailed experimental evidence is lacking so far and the phenotypic appearance of affected probands is not distinguishably different to other patients with IGF-1R mutations.

Phenotype of Patients with IGF1R Mutations

Like patients with IGF-1 mutations, patients with IGF-1R mutations are born small for gestational age and fail to catch up growth in the following years. Moreover, since IGF-1R mutations are assumed to have IGF-1 resistance, affected children have this growth failure despite normal or even elevated IGF-1 serum levels. Birth weight or length is basically ≤–2.0 SDS but height after 1 year usually decreases (≤–2.5 SDS) although thereafter growth might parallel the 3rd percentile during childhood and adolescence. IGF-1 serum levels are relatively high considering the short stature of the patients (usually in the upper normal range). This seems logical for a state of IGF-1 resistance; however, if systematic evaluation is performed in all genetically affected members of a family with IGF-1R mutations, IGF-1 serum levels can be found as low as –0.5 SD below the mean for age adjusted IGF-1 serum levels. This is why we suggest an algorithm for the genetic evaluation of SGA patients without catch-up growth as depicted in figure 2.

Microcephaly has been reported on several occasions, but the growth retardation of the skull seems proportionate in relation to that of the rest of the body. Bone age is delayed in all patients but usually not as severe as in patients with complete IGHD.

Usually, the growth retardation in IGF-1R patients is not accompanied by other syndromic features although some minor morphological abnormalities have been reported for individual cases. In addition to growth retardation, carriers of IGF-1R mutations may present with other comorbidities to a variable degree; however, both

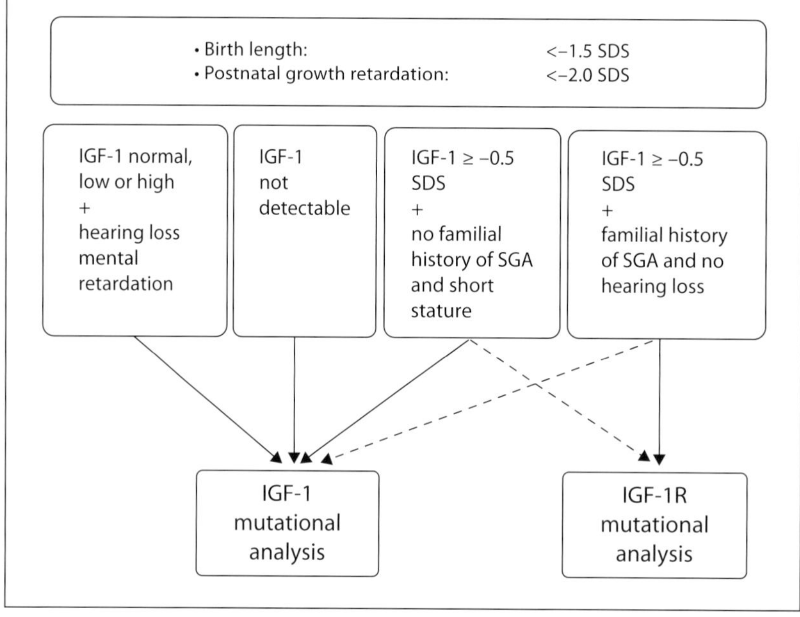

Fig. 2. Suggested algorithm to select patients with SGA and postnatal growth retardation for genetic screening in the IGF-1 and/or IGF-1R genes. Patients with 'elevated' IGF-1 levels (present cut-off >−0.5 SDS) but without a family of history would be more likely to have the recessive forms of IGF-1 gene defects (solid arrow); however, an IGF-1R defect cannot be excluded (dashed arrow). In the dominantly inherited forms of IGF-1R defects, a family history would be more likely.

the frequency and the severity of these comorbidities are not clear at present. In approximately half of the cases reported different degrees of psychomotor retardation have been observed. Patients with a complete loss of bioactive IGF-1 are mentally retarded and they usually have sensorineural deafness.

When IGF-1R patients receive a formal neuropsychological evaluation the results are somewhat ambiguous: the girl reported by Abuzzahab et al. [3] revealed a verbal IQ of 134 and a performance IQ of 89. Another patient reported by Wallborn et al. [21] demonstrated developmental retardation (−2.0 SDS in the Kaufman Assessment Battery for Children), whereas a third patient reported by Kawashima et al. [17] showed mental retardation with an IQ of 60 (Wechsler Intelligence Scale for Children). However, in other patients with IGF-1R mutations intelligence and school performance were reported to be above average.

Both homo- and heterozygous carriers of an inactivating IGF-1 mutation only had a mildly disturbed glucose tolerance, if at all [6]. Within the family of the Gly1155Ala IGF-1R patient, one mutational carrier and six additional relatives, for which genotyping data were not available, developed overt type 1 or 2 diabetes mellitus. Symptomatic diabetes mellitus was also diagnosed as early as age 14.5 years in the index patient bearing the Glu121Lys/Glu234Lys compound heterozygous mutation. In a family carrying the Tyr387X mutation four members displayed alterations in carbohydrate metabolism, characterized by a variable phenotype ranging from normal glucose tolerance in the presence of insulin resistance to impaired glucose tolerance and fasting hyperglycemia in association with both insulin resistance and disturbed β-cell function. However, one has to realize that because genetic and nongenetic factors partially overlap in determining susceptibility for diabetes, it remains difficult to estimate the individual contribution of each IGF-1R mutation.

Treatment of Short Stature in IGF-1R Patients

Most of the patients with IGF-1R defects have received rhGH treatment because they failed to show catch-up growth after being born small for gestational age. GH treatment of SGA children in general seems somewhat less effective compared to patients with GH deficiency even at higher doses. However, the growth response of patients with IGF-1R mutations seems to be below the average of a comparable SGA group without IGF-1R mutations [25]. Due to the fact that the number of documented SGA patients with IGF-1R mutations are limited and that physicians usually tend to lower hGH dosages when they encounter elevated IGF-1 serum levels in patients treated with rhGH, at present it is still difficult to draw a final conclusion in respect to the efficacy of rhGH treatment in these children.

Conclusions

At the lower end of the GHRH-IGF-1 axis genetic defects can cause states of GH resistance due to IGF-1 deficiency or IGF-1 resistance, which are both characterized by intrauterine and postnatal growth failure. Whereas IGF-1 defects seem to be rare and the phenotype is more profound, ongoing research might prove that this disorder may be more variable than realized at present.

In contrast, mutations within the IGF-1R are more common among patients with IUGR possibly accounting for up to 15% of all SGA patients who show 'normal' or elevated IGF-1 serum levels. The phenotype is the result of a functional haploinsufficiency of the IGF-1R and therefore it seems to be variable. Because onset of associated co-morbidities might occur late in adult life, longitudinal surveillance of affected patients employing appropriate diagnostic and analytical methods is required. Future research therefore might show whether these patients require special attention to health issues in adulthood and whether they require a different treatment protocol than SGA patients in general.

References

1 Nissley SP, Rechler MM: Somatomedin/insulin-like growth factor tissue receptors. Clin Endocrinol Metab 1984;13:43–67.

2 Ullrich A, Gray A, Tam AW, Yang-Feng T, Tsubokawa M, Collins C, Henzel W, Le Bon T, Kathuria S, Chen E, et al: Insulin-like growth factor I receptor primary structure: comparison with insulin receptor suggests structural determinants that define functional specificity. EMBO J 1986;5: 2503–2512.

3 Abuzzahab MJ, Schneider A, Goddard A, Grigorescu F, Lautier C, Keller E, Kiess W, Klammt J, Kratzsch J, Osgood D, Pfäffle R, Raile K, Seidel B, Smith RJ, Chernausek SD: IGF-I receptor mutations resulting in intrauterine and postnatal growth retardation. N Engl J Med 2003;349:2211–2222.

4 Liu J-P, Baker J, Perkins AS, Robertson EJ, Efstratiadis A: Mice carrying null mutations of the genes encoding insulin-like growth factor I (Igf-1) and type 1 IGF receptor (Igf1r). Cell 1993;75: 59–72.

5 Woods KA, Camacho-Hubner C, Savage MO, Clark AJ: Intrauterine growth retardation and postnatal growth failure associated with deletion of the insulin-like growth factor I gene. N Engl J Med 1996;335:1363–1367.

6 Walenkamp MJ, Karperien M, Pereira AM, Hilhorst-Hofstee Y, van Doorn J, Chen JW, Mohan S, Denley A, Forbes B, van Duyvenvoorde HA, van Thiel SW, Sluimers CA, Bax JJ, de Laat JA, Breuning MB, Romijn JA, Wit JM: Homozygous and hetero-zygous expression of a novel insulin-like growth factor-I mutation. J Clin Endocrinol Metab 2005;90: 2855–2864.

7 Woods KA, Camacho-Hubner C, Barter D, Clark AJ, Savage MO: Insulin-like growth factor I gene deletion causing intrauterine growth retardation and severe short stature. Acta Paediatr Suppl 1997; 423:39–45.

8 Netchine I, Azzi S, Houang M, Seurin D, Perin L, Ricort JM, Daubas C, Legay C, Mester J, Herich R, Godeau F, Le Bouc Y: Partial primary deficiency of insulin-like growth factor (IGF)-I activity associ-ated with IGF-1 mutation demonstrates its critical role in growth and brain development. J Clin Endocrinol Metab 2009;94:3913–3921.

9 Lee EK, Gorospe M: Posttranscriptional regulation of the insulin and insulin-like growth factor sys-tems. Endocrinology 2010;151:1403–1408.

10 Adams TE, Epa VC, Garrett TP, Ward CW: Structure and function of the type 1 insulin-like growth factor receptor. Cell Mol Life Sci 2000;57:1050–1093.

11 Belfiore A, Frasca F, Pandini G, Sciacca L, Vigneri R: Insulin receptor isoforms and insulin receptor/insulin-like growth factor receptor hybrids in physi-ology and disease. Endocr Rev 2009;30:586–623.

12 Choi JH, Kang M, Kim GH, Hong M, Jin HY, Lee BH, Park JY, Lee SM, Seo EJ, Yoo HW: Clinical and functional characteristics of a novel heterozygous mutation of the IGF-1R gene and IGF-1R haploin-sufficiency due to terminal 15q26.2->qter deletion in patients with intrauterine growth retardation and postnatal catch-up growth failure. J Clin Endocrinol Metab 2011;96:E130–E134.

13 Ester WA, van Duyvenvoorde HA, de Wit CC, Broekman AJ, Ruivenkamp CA, Govaerts LC, Wit JM, Hokken-Koelega AC, Losekoot M: Two short children born small for gestational age with insulin-like growth factor 1 receptor haploinsufficiency illustrate the heterogeneity of its phenotype. J Clin Endocrinol Metab 2009;94:4717–4727.

14 Fang P, Schwartz ID, Johnson BD, Derr MA, Roberts CT, Hwa V, Rosenfeld RG: Familial short stature caused by haploinsufficiency of the insulin-like growth factor I receptor due to nonsense-mediated messenger ribonucleic acid decay. J Clin Endocrinol Metab 2009;94:1740–1747.

15 Inagaki K, Tiulpakov A, Rubtsov P, Sverdlova P, Peterkova V, Yakar S, Terekhov S, LeRoith D: A familial insulin-like growth factor-I receptor mutant leads to short stature: clinical and biochemical characterization. J Clin Endocrinol Metab 2007;92: 1542–1548.

16 Kawashima Y, Higaki K, Fukushima T, Hakuno F, Nagaishi JI, Hanaki K, Nanba E, Takahashi SI, Kanzaki S: Novel missense mutation in the IGF-I receptor L2 domain results in intrauterine and post-natal growth retardation. Clin Endocrinol (Oxf) 2012;77:246–254.

17 Kawashima Y, Kanzaki S, Yang F, Kinoshita T, Hanaki K, Nagaishi J-I, Ohtsuka Y, Hisatome I, Ninomoya H, Nanba E, Fukushima T, Takahashi S-I: Mutation at cleavage site of insulin-like growth factor receptor in a short-stature child born with intrauterine growth retardation. J Clin Endocrinol Metab 2005;90:4679–4687.

18 Kruis T, Klammt J, Galli-Tsinopoulou A, Wallborn T, Schlicke M, Müller E, Kratzsch J, Körner A, Odeh R, Kiess W, Pfäffle R: Heterozygous mutation within a kinase-conserved motif of the insulin-like growth factor I receptor causes intrauterine and postnatal growth retardation. J Clin Endocrinol Metab 2010; 95:1137–1142.

19 Mohn A, Marcovecchio ML, de Giorgis T, Pfaeffle R, Chiarelli F, Kiess W: An insulin-like growth factor-I receptor defect associated with short stature and impaired carbohydrate homeostasis in an Italian pedigree. Horm Res Paediatr 2011;76: 136–143.

20 Walenkamp MJ, van der Kamp HJ, Pereira AM, Kant SG, van Duyvenvoorde HA, Kruithof MF, Breuning MH, Romijn JA, Karperien M, Wit JM: A variable degree of intrauterine and postnatal growth retardation in a family with a missense mutation in the insulin-like growth factor I receptor. J Clin Endocrinol Metab 2006;91:3062–3070.

21 Wallborn T, Wüller S, Klammt J, Kruis T, Kratzsch J, Schmidt G, Schlicke M, Müller E, Schmitz van de Leur H, Kiess W, Pfäffle R: A heterozygous mutation of the insulin-like growth factor-I receptor causes retention of the nascent protein in the endoplasmic reticulum and results in intrauterine and postnatal growth retardation. J Clin Endocrinol Metab 2010;95:2316–2324.

22 Fang P, Cho YH, Derr MA, Rosenfeld RG, Hwa V, Cowell CT: Severe short stature caused by novel compound heterozygous mutations of the insulin-like growth factor 1 receptor (IGF-1R). J Clin Endocrinol Metab 2012;97:E243–E247.

23 Vashisth H, Abrams CF: All-atom structural models for complexes of insulin-like growth factors IGF-1 and IGF2 with their cognate receptor. J Mol Biol 2010;400:645–658.

24 Pautsch A, Zoephel A, Ahorn H, Spevak W, Hauptmann R, Nar H: Crystal structure of bisphosphorylated IGF-1 receptor kinase: insight into domain movements upon kinase activation. Structure 2001;9:955–965.

25 Klammt J, Kiess W, Pfaffle R: IGF-1R mutations as cause of SGA. Best Pract Res Clin Endocrinol Metab 2011;25:191–206.

Prof. Dr. med. Roland Pfäffle
Department of Pediatrics, University of Leipzig Medical School
Liebigstrasse 20a
DE–04103 Leipzig (Germany)
Tel. +49 341 9726330, E-Mail rpfaeffle@medizin.uni-leipzig.de

Mullis P-E (ed): Developmental Biology of GH Secretion, Growth and Treatment.
Endocr Dev. Basel, Karger, 2012, vol 23, pp 52–59 (DOI: 10.1159/000341748)

New Detection Methods of Growth Hormone and Growth Factors

Martin Bidlingmaier

Endocrine Research Laboratories, Medizinische Klinik und Poliklinik IV, Ludwig Maximilians University, Munich, Germany

Abstract

Human growth hormone (GH), but also GH related growth factors like the insulin-like growth factor-1 (IGF-1) are known to be abused in sports. Although the scientific evidence supporting a distinct effect of GH on performance in healthy trained subjects is limited, it has been repeatedly found with athletes or trainers, and the recent introduction of a first test to detect GH doping has led to a number of positive cases. Currently, there is no test for the detection of IGF-1 introduced worldwide, but confiscation of the drug from sports teams can be taken as indirect evidence for its abuse. The major biochemical difficulty for the detection of GH is that the recombinant form is identical in physicochemical properties to the endogenous GH secreted by the pituitary gland. Furthermore, the very short half-life of GH in circulation inherently shortens the window of opportunity where the drug can be detected. Two strategies have been followed for more than a decade to develop a test to detect the application of recombinant GH: the marker approach, which is based on the elevation of GH-dependent markers above the level seen under physiological conditions evoked by administration of recombinant GH, and the isoform approach, which is based on a change in the pattern of GH isoforms in circulation following the injection of recombinant GH.

Evidence for Performance Enhancement and Abuse of Recombinant GH in Sports

Human growth hormone (GH) is a protein which is secreted by the pituitary gland in a pulsatile manner. Beyond its well-known effects on longitudinal growth during childhood and adolescence, GH remains a very important metabolic hormone across the lifetime. One of the most frequently reported and best documented effects in healthy adults is the decrease in fat mass. One can speculate if – and if yes, in which sports disciplines – this effect can be of advantage for performance. But also beyond athletics, the lipolytic effect of GH is frequently used to advertise GH as a lifestyle hormone. Already a decade ago had been reported that the use of GH for anti-aging and for athletic enhancement accounts for approximately 30% of GH prescriptions in the United States [1]. There were also a couple of surveys among body builders and high school or university students confirming that in spite of the high costs recombinant GH, and more recently also IGF-1, are being abused [2–4].

Within the scientific community, there is ongoing debate whether or not administration of GH to healthy trained subjects actually has performance-enhancing effects. A lot of the claims are simply derived from extrapolations of effects seen in GH-deficient adults. Obviously, the effects in severely GH-deficient adults – which commonly have a decreased muscle mass and an increased fat mass – can very well be different from the effects one could expect in trained healthy subjects. It has been speculated if the uncritical repetition of effects which theoretically can be attributed to GH (but have never been proven in placebo-controlled trials) has contributed to the popularity of GH in sports [5]. On the other hand, it is important to keep in mind that in the case of testosterone – a potent anabolic drug abused for several decades, where nobody really had questioned its performance-enhancing effect in healthy subjects –, it was only in 1996 when the first controlled study scientifically proving this effect was published [6]. Scientists and physicians have to accept and understand that frequently athletes' 'knowledge' about potential effects can come years before the recognition of these effects in clinical medicine. In the case of GH, it was before the introduction of recombinant GH treatment in adults that the 'Underground steroid handbook' had already promoted the beneficial effects of GH on performance [7, 8]! A few years ago, systematic reviews of the effects of supraphysiological doses of GH in healthy elderly subjects [9] and in sports [10] revealed that despite the lipolytic effect it is difficult to identify convincing evidence from controlled studies supporting the notion of major benefits. Especially, there is no demonstration of any effects on strength. Interestingly, a very recent, well-controlled trial in recreational athletes [11] could demonstrate that growth hormone supplementation – in addition to its influence on body composition – increased sprint capacity when administered alone and in combination with testosterone. The same study, however, confirmed the absence of effects on VO_{2max} and various indicators of strength (dead lift and jump height). Overall, the effects of GH on athletic performance remain unclear, with several potential mechanisms being discussed [12]. One should also not forget that a small advantage an individual athlete might take from any performance-enhancing drug might be very relevant in terms of winning a competition or not – although the order of magnitude might not be significant when investigated in a scientific study!

How Can the Application of Growth Hormone Be Detected?

Recombinant human GH is a 191 amino acid protein which for pharmaceutical purposes usually is produced in *Escherichia coli*. Endogenous GH has no O-linked glycosylation sites, and the preparations of recombinant GH available for treatment of patients are identical in amino acid sequence and in other physicochemical to the main isoform of GH produced by the pituitary gland. This main isoform has a molecular weight of 22 kD. Once injected, each GH molecule of recombinant origin becomes indistinguishable from the 22-kD GH molecules derived from the bodies' own pituitary secretion. This identity makes it particularly challenging to develop a test method able to prove the illicit use of recombinant GH in doping tests. Other difficulties include the very short half-life of GH in the circulation [13], the very low and variable GH concentrations in urine (excluding the use of this most common matrix in doping tests for analysis of GH [14]) and the lack of established mass spectrometric approaches to measure GH in complex biological samples.

To overcome the difficulties, two independent approaches have been followed during the last 15 years: one approach – meanwhile termed the 'marker approach' – makes use of the fact that several pharmacodynamic markers of GH action change much more after the administration of recombinant GH than after normal physiological stimulation of endogenous GH release [15]. The other approach is known as the 'isoform approach' and is based upon the fact that endogenous GH in the circulation represents a group of closely related but not identical molecular isoforms, while recombinant GH consists of the 22-kD monomeric isoform only. Injection of the recombinant 22-kD GH isoform leads to a suppression of pituitary GH secretion through negative feedback mechanisms, and thereby to a reduction in isoforms other than 22 kD in the circulation. This change in the isoform spectrum present in serum can be measured and is used to demonstrate that recombinant GH had been injected [16].

The Marker Approach

Among the pharmacodynamic markers of GH action, especially IGF-1 [17], and markers of bone and collagen turnover [18] have been shown to exhibit dramatic changes in concentration after the administration of recombinant GH. Most importantly, the changes seen after the administration of recombinant GH clearly exceeded natural fluctuation [19] or changes seen after exercise or injury [20]. The best sensitivity and specificity was found when two markers, namely IGF-1 and P-III-P, were combined using a formula (so-called 'discriminant functions') which also includes age and sex of the athlete [21]. A major advantage of this approach is that the pharmacodynamic markers of GH action have a much longer half-life than GH itself, making a longer window of opportunity for detection very likely.

The approach has been validated in athletes from different ethnic groups [22]. A crucial point for the validity of the test is the consistency of the assay results in terms of assay method, within-subject variability [23], as well as pre-analytical stability of the analytes [24]. Multicentric studies involving several laboratories accredited by the World Anti-Doping Agency (WADA) for conducting doping tests are underway, and a worldwide implementation of the test method is expected in the near future.

The Isoform Approach

Human GH, as secreted by the pituitary gland, is not a homogeneous substance. In contrast, it consists of a spectrum of molecular isoforms. Next to the major isoform, the 191 amino acid protein with a molecular weight of 22 kD, the smaller 20-kD isoform of GH lacking amino acids 32–46, is the second most abundant isoform [25]. Further variability is introduced because GH isoforms also exist as homo- or heterodimers and multimers. Finally, desamido- and acylated forms of GH exist also. Importantly, the relative abundance of the different isoforms – although exhibiting some variability between individuals – remains constant under conditions where GH release from the pituitary is stimulated. There is a long list of studies using many different techniques to measure different GH isoforms, and all these studies clearly demonstrate that the different GH isoforms are secreted in parallel [26].

In contrast to the complex spectrum of molecular isoforms, the recombinant GH used for therapeutic purposes (and also abused for doping) consists of the monomeric 22-kD isoform only. Following the injection of recombinant GH, the pituitary gland stops the secretion of the wide spectrum of molecular isoforms due to negative feedback, and the (administered exogenously) 22-kD isoform becomes predominant in the circulation. It was shown more than 10 years ago that the change in the isoform composition of GH following injection of recombinant GH can be measured in serum samples using specific immunoassays based on monoclonal antibodies with different affinities for the different isoforms [27]. One of the assays, the 'rec assay', exhibits a high preference to measure the 22-kD monomeric GH, whereas the other assay, named the 'pit assay', binds a broader spectrum of mono-, di- and multimeric GH isoforms. A 'rec/pit ratio' is calculated from both assay results, reflecting the relative abundance of the respective isoforms. The rec/pit ratio exhibits some degree of variability between individuals, reflecting the degree of variability in isoforms. However, the rec/pit ratios seen after the administration of recombinant GH, when almost all the GH in the circulation is attributable to the 22-kD isoform, by far exceed the range of variability seen in 'clean' subjects. Therefore, a rec/pit ratio above a certain threshold can be used as an indication for a preceding administration of recombinant GH. The efficacy of this testing principle to detect the administration of recombinant GH had been already shown in the initial studies. However, it took several years from the first description to develop the assays needed according to the requirements established by WADA for approving a test method: two independent sets of assays involving antibodies recognizing different epitopes had to be developed, and the assays had to be thoroughly validated [28]. Furthermore, the assays had to be implemented worldwide in WADA-accredited laboratories [29], and several studies had to be conducted to establish decision limits allowing declaring a sample 'positive' or – to use the more correct term – declare a results an 'adverse analytical finding'. Studies from Japan [30] and China [31] have shown the efficacy as well as the inherent limitations in terms of the window of opportunity. The chance to catch the cheaters is highest within the first 24 h after injection, and rapidly drops thereafter. This short window of detection was expected as the pituitary gland starts secretion of the endogenous isoforms within 1 day after the injection of recombinant GH [32], leading to a rapid drop in the rec/pit ratio.

After validation of the test kits by accredited laboratories, WADA published a Guideline of hGH Isoform Differential Immunoassays for Anti-Doping Analyses (version 1.0) in June 2010 [33], which forms the basis for a more widespread use of the test. Despite the short window of opportunity to detect the preceding administration of recombinant GH, several athletes have been tested positive since WADA-accredited laboratories started to use the test more frequently in 2008. Only in April 2012, did the US Anti-Doping Agency announce that a professional weightlifter had accepted a 2-year ban following a positive test [34]. The detection of GH abuse therefore is possible by the isoform approach, but intelligent testing strategies including a significant number of unannounced out-of-competition tests is a prerequisite for the deterrent effect of the test method.

Conclusion

After many years of research involving different teams in many countries it became possible to detect GH doping in sports. Obviously, the likelihood of being caught still seems to be low. The inherent limitation of the currently used isoform test is related to the fact that this test relies on the physiological suppression of pituitary GH secretion after injection of recombinant GH, which lasts only for a limited time period. As soon as pituitary GH secretion is recurring, the rec/pit ratios decline, leading to negative test results. In line with this, it was demonstrated recently that the concomitant use of GH secretagogues prevents the detection of recombinant GH administration [35]. Most of the GH secretagogues, however, can be detected by antidoping laboratories, putting athletes who try to mask GH abuse at risk for being banned for the illicit use of secretagogues. Generally, the marker approach bears the promise of a longer window of opportunity. Generally, this approach seems to be more sensitive in males as compared to females, and the upcoming implementation of the test in several different anti-doping laboratories will finally demonstrate the efficacy.

Both approaches are currently using classical immunoassays to measure the concentrations of the analytes in question. If in the future mass-based approaches like liquid chromatography tandem mass spectrometry (LC-MS/MS) will be sensitive enough to replace or at least complement the immunoassays used is an open question. Measurement of GH by LC-MS/MS is possible, but currently the technique is not sensitive enough to allow application in the low concentration ranges frequently seen in humans [36]. Furthermore, even if the measurement of 22-kD GH becomes possible with high sensitivity in serum samples, the detection of recombinant GH still remains challenging. Adoption of the isoform approach to mass spectrometry would also require the availability of a measurement technique for other (pituitary derived) isoforms. In case of the marker approach, the major progress has been made regarding the measurement of IGF-1 by mass spectrometry [37, 38]. However, similar progress has not been made for the analysis of P-III-P concentrations.

Whatever test might be used in the future to detect GH doping, it remains important to keep in mind that the prevention of GH abuse is not only required because its use might provide an unfair advantage for cheating athletes. It is also important to send the message to athletes, trainers as well as the public that the use of GH outside well-defined approved indications bears the risk of serious side effects. Off-label use of GH in otherwise healthy subjects for whatever purpose is of questionable efficacy, but excess GH definitely harms.

References

1   Vance ML: Can growth hormone prevent aging? N Engl J Med 2003;348:779–780.
2   Calfee R, Fadale P: Popular ergogenic drugs and supplements in young athletes. Pediatrics 2006;117: e577–e589.
3   Holt RI, Erotokritou-Mulligan I, Sonksen PH: The history of doping and growth hormone abuse in sport. Growth Horm IGF Res 2009;19:320–326.
4   Brennan BP, Kanayama G, Hudson JI, Pope HG Jr: Human growth hormone abuse in male weightlifters. Am J Addict 2011;20:9–13.
5   Rennie MJ: Claims for the anabolic effects of growth hormone: a case of the emperor's new clothes? Br J Sports Med 2003;37:100–105.
6   Bhasin S, Storer TW, Berman N, Callegari C, Clevenger B, Phillips J, Bunnell TJ, Tricker R, Shirazi A, Casaburi R: The effects of supraphysiologic doses of testosterone on muscle size and strength in normal men. N Engl J Med 1996;335:1–7.
7   Sonksen PH: Insulin, growth hormone and sport. J Endocrinol 2001;170:13–25.
8   Nelson AE, Ho KK: Abuse of growth hormone by athletes. Nat Clin Pract Endocrinol Metab 2007;3: 198–199.
9   Liu H, Bravata DM, Olkin I, Nayak S, Roberts B, Garber AM, Hoffman AR: Systematic review: the safety and efficacy of growth hormone in the healthy elderly. Ann Intern Med 2007;146:104–115.
10 Liu H, Bravata DM, Olkin I, Friedlander A, Liu V, Roberts B, Bendavid E, Saynina O, Salpeter SR, Garber AM, Hoffman AR: Systematic review: the effects of growth hormone on athletic performance. Ann Intern Med 2008;148:747–758.
11 Meinhardt U, Nelson AE, Hansen JL, Birzniece V, Clifford D, Leung KC, Graham K, Ho KK: The effects of growth hormone on body composition and physical performance in recreational athletes: a randomized trial. Ann Intern Med 2010;152: 568–577.
12 Birzniece V, Nelson AE, Ho KK: Growth hormone and physical performance. Trends Endocrinol Metab 2011;22:171–178.
13 Holl RW, Schwarz U, Schauwecker P, Benz R, Veldhuis JD, Heinze E: Diurnal variation in the elimination rate of human growth hormone (GH): the half-life of serum GH is prolonged in the evening, and affected by the source of the hormone, as well as by body size and serum estradiol. J Clin Endocrinol Metab 1993;77:216–220.
14 Saugy M, Cardis C, Schweizer C, Veuthey JL, Rivier L: Detection of human growth hormone doping in urine: out of competition tests are necessary. J Chromatogr B Biomed Appl 1996;687:201–211.
15 Holt RI, Erotokritou-Mulligan I, McHugh C, Bassett EE, Bartlett C, Fityan A, Bacon JL, Cowan DA, Sonksen PH: The GH-2004 project: the response of IGF1 and type III pro-collagen to the administration of exogenous GH in non-Caucasian amateur athletes. Eur J Endocrinol 2010;163:45–54.
16 Bidlingmaier M, Strasburger CJ: Technology insight: detecting growth hormone abuse in athletes. Nat Clin Pract Endocrinol Metab 2007;3:769–777.

17 Wallace JD, Cuneo RC, Baxter R, Orskov H, Keay N, Pentecost C, Dall R, Rosen T, Jorgensen JO, Cittadini A, Longobardi S, Sacca L, Christiansen JS, Bengtsson BA, Sonksen PH: Responses of the growth hormone (GH) and insulin-like growth factor axis to exercise, GH administration, and GH withdrawal in trained adult males: a potential test for GH abuse in sport. J Clin Endocrinol Metab 1999;84:3591–3601.

18 Longobardi S, Keay N, Ehrnborg C, Cittadini A, Rosen T, Dall R, Boroujerdi MA, Bassett EE, Healy ML, Pentecost C, Wallace JD, Powrie J, Jorgensen JO, Sacca L: Growth hormone (GH) effects on bone and collagen turnover in healthy adults and its potential as a marker of GH abuse in sports: a double blind, placebo-controlled study. The GH-2000 Study Group. J Clin Endocrinol Metab 2000;85: 1505–1512.

19 Healy ML, Dall R, Gibney J, Bassett E, Ehrnborg C, Pentecost C, Rosen T, Cittadini A, Baxter RC, Sonksen PH: Toward the development of a test for growth hormone (GH) abuse: a study of extreme physiological ranges of GH-dependent markers in 813 elite athletes in the postcompetition setting. J Clin Endocrinol Metab 2005;90:641–649.

20 Erotokritou-Mulligan I, Bassett EE, Bartlett C, Cowan D, McHugh C, Seah R, Curtis B, Wells V, Harrison K, Sonksen PH, Holt RI: The effect of sports injury on insulin-like growth factor-I and type 3 procollagen: implications for detection of growth hormone abuse in athletes. J Clin Endocrinol Metab 2008;93:2760–2763.

21 Erotokritou-Mulligan I, Bassett EE, Kniess A, Sonksen PH, Holt RI: Validation of the growth hormone (GH)-dependent marker method of detecting GH abuse in sport through the use of independent data sets. Growth Horm IGF Res 2007;17:416–423.

22 Erotokritou-Mulligan I, Bassett EE, Cowan DA, Bartlett C, McHugh C, Sonksen PH, Holt RI: Influence of ethnicity on IGF-1 and procollagen III peptide (P-III-P) in elite athletes and its effect on the ability to detect GH abuse. Clin Endocrinol (Oxf) 2009;70:161–168.

23 Nguyen TV, Nelson AE, Howe CJ, Seibel MJ, Baxter RC, Handelsman DJ, Kazlauskas R, Ho KK: Within-subject variability and analytic imprecision of insulin-like growth factor axis and collagen markers: implications for clinical diagnosis and doping tests. Clin Chem 2008;54:1268–1276.

24 Guha N, Erotokritou-Mulligan I, Bartlett C, D AC, Bassett EE, Stow M, Sonksen PH, Holt RI: The effects of a freeze-thaw cycle and pre-analytical storage temperature on the stability of insulin-like growth factor-I and pro-collagen type III N-terminal propeptide concentrations: implications for the detection of growth hormone misuse in athletes. Drug Test Anal 2012;4:455–459.

25 Baumann G: Growth hormone heterogeneity: genes, isohormones, variants, and binding proteins. Endocr Rev 1991;12:424–449.

26 Baumann GP: Growth hormone doping in sports: a critical review of use and detection strategies. Endocr Rev 2012;33:155–186.

27 Wu Z, Bidlingmaier M, Dall R, Strasburger CJ: Detection of doping with human growth hormone. Lancet 1999;353:895.

28 Bidlingmaier M, Suhr J, Ernst A, Wu Z, Keller A, Strasburger CJ, Bergmann A: High-sensitivity chemiluminescence immunoassays for detection of growth hormone doping in sports. Clin Chem 2009; 55:445–453.

29 Barroso O, Schamasch P, Rabin O: Detection of GH abuse in sport: Past, present and future. Growth Horm IGF Res 2009;19:369–374.

30 Okano M, Nishitani Y, Sato M, Kageyama S: Effectiveness of GH isoform differential immunoassay for detecting rhGH doping on application of various growth factors. Drug Test Anal 2012 Jun 25. doi: 10.1002/dta.1385 [Epub ahead of print].

31 Jing J, Yang S, Zhou X, He C, Zhang L, Xu Y, Xie M, Yan Y, Su H, Wu M: Detection of doping with rhGH: excretion study with WADA-approved kits. Drug Test Anal 2011;3:784–790.

32 Keller A, Wu Z, Kratzsch J, Keller E, Blum WF, Kniess A, Preiss R, Teichert J, Strasburger CJ, Bidlingmaier M: Pharmacokinetics and pharmacodynamics of GH: dependence on route and dosage of administration. Eur J Endocrinol 2007;156: 647–653.

33 WADA 2010 Detection of Doping with Human Growth Hormone. Available at: http://www.wada-ama.org/en/World-Anti-Doping-Program/Sports-and-Anti-Doping-Organizations/Model-Rules–Guidelines/Guidelines/ [June 16, 2012].

34 USADA 2012 US Weightlifting Athlete, Mendes, Accepts Sanction for Anti-Doping Rule Violation http://www.usada.org/media/sanction-mendes 4162012 [June 16, 2012].

35 Okano M, Nishitani Y, Sato M, Ikekita A, Kageyama S: Influence of intravenous administration of growth hormone releasing peptide-2 (GHRP-2) on detection of growth hormone doping: growth hormone isoform profiles in Japanese male subjects. Drug Test Anal 2010;2:548–556.

36 Arsene CG, Henrion A, Diekmann N, Manolopoulou J, Bidlingmaier M: Quantification of growth hormone in serum by isotope dilution mass spectrometry. Anal Biochem 2010;401:228–235.

37 Bredehoft M, Schanzer W, Thevis M: Quantification of human insulin-like growth factor-1 and qualitative detection of its analogues in plasma using liquid chromatography/electrospray ionisation tandem mass spectrometry. Rapid Commun Mass Spectrom 2008;22:477–485.

38 Thomas A, Kohler M, Schanzer W, Delahaut P, Thevis M: Determination of IGF-1 and IGF-2, their degradation products and synthetic analogues in urine by LC-MS/MS. Analyst 2011;136:1003–1012.

Martin Bidlingmaier, MD
Medizinische Klinik und Poliklinik IV, Klinikum der Ludwig-Maximilians-University
Ziemssenstrasse 1
DE–80336 Munich (Germany)
Tel. +49 89 5160 2277, E-Mail martin.bidlingmaier@med.uni-muenchen.de

Mullis P-E (ed): Developmental Biology of GH Secretion, Growth and Treatment.
Endocr Dev. Basel, Karger, 2012, vol 23, pp 60–70 (DOI: 10.1159/000341750)

Imprinted Anomalies in Fetal and Childhood Growth Disorders: The Model of Russell-Silver and Beckwith-Wiedemann Syndromes

Irène Netchine · Sylvie Rossignol · Salah Azzi · Fréderic Brioude · Yves Le Bouc

Laboratoire d'Explorations Fonctionnelles Endocriniennes, APHP, Hôpital Armand Trousseau, INSERM UMRS-938 team 4, Université Pierre et Marie Curie-Paris 6, Paris, France

Abstract

Fetal growth is a complex process. Its restriction is associated with morbidity and long term metabolic consequences. Imprinted genes have a critical role in mammalian fetal growth. The human chromosome 11p15 encompasses two imprinted domains regulated by their own differentially methylated region (DMR), also called Imprinted Control Region (ICR1 at the *H19/IGF-2* domain, paternally methylated), and ICR2 at the *KCNQ1/CDKN1C* domain (maternally methylated). Loss of imprinting at these two domains is implicated in two growth disorders clinically opposite. A loss of DNA methylation (LOM) at ICR1 is identified in over 50% of patients with Russell-Silver syndrome (RSS), characterized by intrauterine and postnatal growth retardation, spared cranial growth, frequent body asymmetry and severe feeding difficulties. Inversely, a gain of methylation at ICR1 is found in 10% of patients with Beckwith-Wiedemann syndrome (BWS), an overgrowth syndrome with an enhanced childhood tumor risk. We have identified over 150 RSS patients with 11p15 LOM allowing long-term follow-up studies and proposal of clinical guidelines. We also found that ~10% of RSS patients and ~25% of BWS patients have multilocus LOM at imprinted regions other than ICR1 or ICR2 11p15, respectively. Recent studies have identified *cis*-acting regulatory elements and *trans*-acting factors involved in the regulation of 11p15 imprinting, establishing new potential mechanisms of RSS and BWS.

Fetal growth is a complex process involving multiple environmental, epigenetic and genetic factors. Intrauterine growth retardation or small for gestational age (IUGR/SGA) represents about 5% of the births and is associated with severe morbidity. Indeed, IUGR patients are exposed to an enhanced risk to develop cardiovascular or metabolic diseases in adult life (fetal programming or the adapting hypothesis, Developmental origins of health and disease) [1]. By contrast, fetal overgrowth syndromes are associated with developmental abnormalities, tissue and organ hyperplasia and a higher risk of childhood tumors [2].

Imprinting Marks

It has been recognized that faithful genetic and epigenetic programming has a critical role in controlling mammalian fetal and postnatal growth. The most studied epigenetic mark is DNA methylation of CpG dinucleotide cytosine residues within gene promoters, transposons and imprinted regions. DNA methylation is usually associated with a repressed chromatin state and gene silencing whereas nonmethylation is associated with an active chromatin state allowing transcription activation. Other key regulators such as histone modifications (acetylation and methylation, affecting the N-terminal tail of histones) play pivotal roles in chromatin organization as well as gene regulation. Genomic imprinting is an epigenetic mechanism whereby expression of a subset of genes is restricted to a single parental allele. Approximately 100 imprinted genes have been identified in mammals (http://www.geneimprint.com) and most of them are organized in clusters throughout the genome and are regulated by an imprinting control region (ICR) [3].

Imprinting Cycle

The methylation of the genome undergoes dynamic reprogramming during fetal development (fig. 1). The first step takes place in primordial germ cells where methylation is erased from the ICRs and further re-established, on cytosine residues of the CpG dinucleotides, according to the gender-specific gamete [4]. The second important widespread epigenetic reprogramming occurs in the pre-implantation period where the whole genome undergoes a wave of general demethylation soon after fertilization, while the imprinting are maintained during this period and then followed by a progressive wave of lineage specific de novo methylation beginning at blastocyste stage. The mechanisms leading to establishment and maintenance of allele-specific DNA methylation at ICRs, throughout fetal development and adulthood, are very complex and not fully understood. They involve not only *cis-* and *trans-*acting regulatory factors [5, 6] but also environmental conditions such as dietary factors or assisted reproductive technology (ART) [7–9]. Loss of imprinting (LOI) through gain (GOM) or loss (LOM) of DNA methylation is implicated in several human diseases and cancer (table 1) [10]. Note that the frequency of LOI as an etiology of these diseases is variable suggesting that some imprinted regions are more resistant than others to LOI.

11p15 Region

The human chromosome 11p15 encompasses two imprinted domains (fig. 2) important in the control of fetal and postnatal growth. Each domain is differentially methylated and regulated by its own ICR (ICR1 at the telomeric region for the *H19/IGF-2* domain which is methylated on the paternal allele, and ICR2 at the centromeric region for the *KCNQ1OT1/CDKN1C* domain which is methylated on the maternal allele). LOI at these two domains is involved in two growth disorders clinically opposite. Indeed, LOM at ICR1 is identified in over 50% of Russell-Silver syndrome patients (RSS, a growth-restriction syndrome) [11, 12] whereas GOM at ICR1 is found in

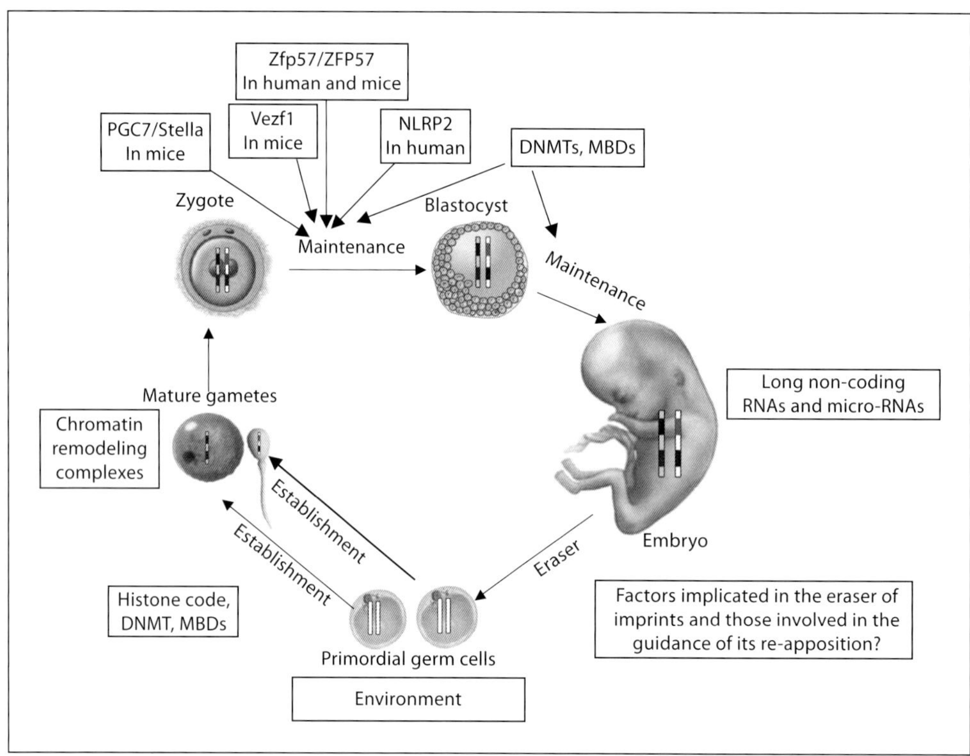

Fig. 1. Illustration of the dynamic imprinting reprogramming during fetal development and the various factors (currently identified) involved in the regulation of this complex process. Some of these factors have been identified in both humans and mice, but it is likely that a huge number of other factors are involved at various steps of this cycle. Adapted from [31].

Table 1. Frequency of LOI involvement in some human syndromes

Pathology	OMIM	Growth characteristic	Epimutation frequency	Chromosome
Transient neonatal diabetes mellitus	601410	IUGR	20%	6q24
Russell-Silver syndrome	180860	IUGR	>50%	11p15
Beckwith-Wiedemann syndrome	130650	macrosomia	70%	11p15
Maternal UPD14-like syndrome	608149	IUGR	6 reported cases of LOM	14q32
Angelman syndrome	105830	variable	<5%	15q11–13
Prader-Willi syndrome	176270	obesity, short stature	1–2%	15q11–13
Pseudohypoparathyroidism 1b	103580	variable	100%	20q13

 Netchine · Rossignol · Azzi · Brioude · Le Bouc

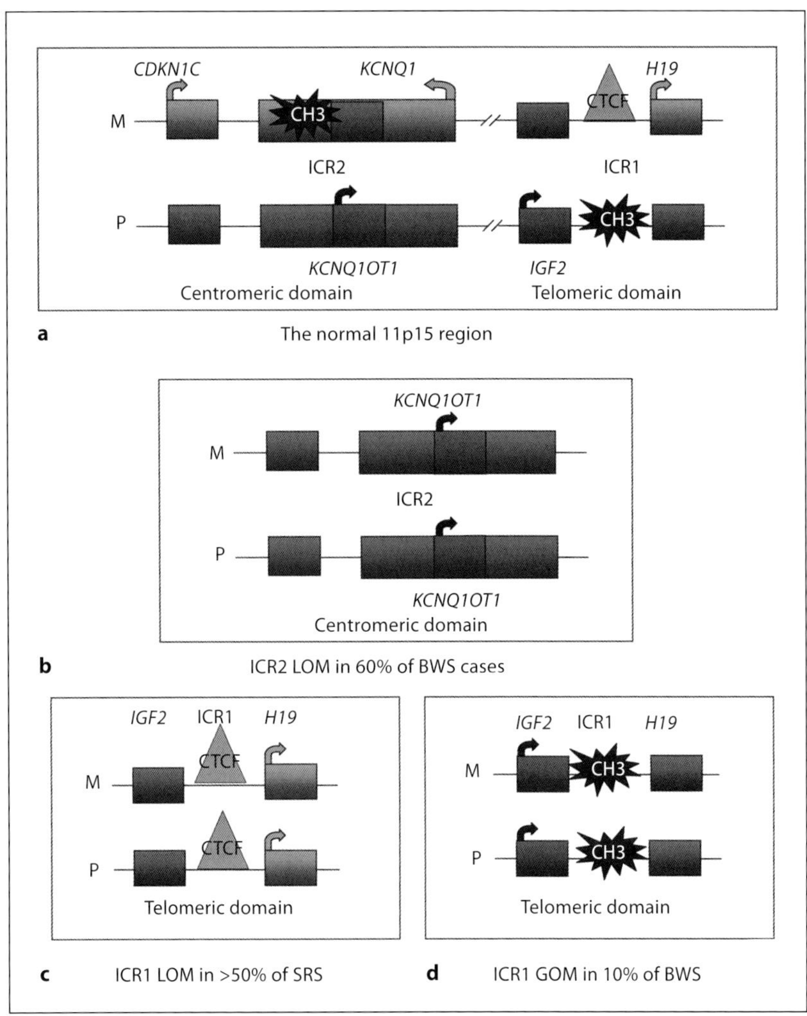

Fig. 2. Schematic representation of the 11p15 region and its imprinting related defects. **a** Normal imprinting of the centromeric and telomeric 11p15 region. ICR1 is methylated on the paternal allele whereas ICR2 is methylated on the maternal allele. These differential methylations at the two ICRs allow monoallelic expression of the different genes of the region. **b** Loss of methylation (LOM) at the ICR2 is found in 60% cases of BWS. In 10% of BWS cases, a gain of methylation at ICR1 is found (**d**). This defect leads to biallelic expression of IGF-2 and loss of expression of H19. By contrast, LOM at ICR1 is found in over 50% of RSS cases leading to biallelic expression of H19 and loss of expression of IGF-2 (**c**). P = Paternal allele; M = maternal allele. Adapted from [31].

10% of Beckwith-Wiedemann syndrome (BWS, an overgrowth syndrome with an increased tumor risk) patients with identified molecular anomalies [13]. This clinical and molecular mirror demonstrates the crucial role of the imprinted *IGF-2* gene in fetal growth. However, the great majority of BWS cases are associated with other molecular defect within the 11p15 region (see later).

Russell-Silver Syndrome

Clinical Diagnosis
- RSS is a clinically heterogeneous syndrome characterized by severe IUGR with
  relative macrocephaly, postnatal growth retardation, a distinctive triangular
  face with prominent forehead and body asymmetry [14–16]. Many other minor
  signs have been added including clinodactyly of the fifth fingers, café-au-lait
  spots, genital abnormalities, hypoglycemia, excessive sweating, blue sclera and
  severe feeding difficulties. No consensus definition has been adopted yet. Recent
  findings in the molecular diagnosis of the syndrome (see below) prompted
  several teams to propose more or less complex clinical scores and to establish
  genotype/phenotype correlations [12, 17–20]. Based on a large cohort of >150
  RSS patients [12, and unpublished data] with an identified molecular defect, we
  proposed a validated clinical scoring system for the diagnosis of RSS:
- the patient must be born SGA (birth weight and/or length ≤–2 SDS (Standard
  Deviation Score) for gestational age),
- and also present at least 3 of the 5 following criteria: (1) postnatal growth
  retardation at 2 years of age or at the nearest measure available, (2) relative
  macrocephaly (i.e. arbitrarily defined when the head circumference at birth is
  at least 1.5 SDS above the birth weight and/or length), (3) body asymmetry, (4)
  prominent forehead, and (5) feeding difficulties during early childhood and/or
  postnatal BMI below –2 SDS at 2 years of age or at the nearest measure available.

Of note, we have included feeding difficulties and/or low BMI as one of the criteria
as they are reported to be particularly frequent and severe in RSS during infancy and
early childhood. This manifestation can be associated with frequent gastrointestinal
disorders, such as constipation, gut dysmobility, gastroesophageal reflux disease [21].
We have retained the prominent forehead as the main characteristic of facial dysmorphy, but it should be assessed before 3 years of age since the characteristic facial features of the RSS may change and are not as marked in late childhood and adulthood
[12, 19].

Molecular Diagnosis
The genetic cause of RSS had remained unknown for a long time. Most of the cases
are sporadic, both genders are equally affected. The first molecular abnormality identified in a significant proportion of patients was maternal UPD for chromosome
7 (mUPD7), present in around 7–10% of the cases [22]. Maternal UPD7 involves
either iso- or heterodisomy of the whole chromosome in most subjects. This finding
implies that one or more genes on this chromosome are imprinted, and that disturbed
imprinting is responsible for the phenotype. So far, two candidate regions on chromosome 7 have been the focus of research: 7p11.1–p14 and 7q31, for which a number of RSS patients with duplications (or inversions) or segmental UPD have been
reported. These regions harbor imprinted genes possibly involved in human growth

Netchine · Rossignol · Azzi · Brioude · Le Bouc

and development, such as *GRB10* and *PEG1/MEST*. However, their implication in the physiopathology of the syndrome is not yet demonstrated.

In 2005, our group described the main molecular defect associated with RSS: ICR1 LOM, present in more than 50% of the cases [11, 12]. In most of the cases, the LOM is partial reflecting the mosaic distribution of the epimutation. In RSS patients, the paternal allele switches to a maternal epigenotype resulting in biallelic expression of *H19* and loss of *IGF-2* expression in pathologic cells (fig. 2). Because these epimutations occur in a mosaic manner and also given the complexity of the imprinting control regions, the molecular diagnosis of these partial methylation anomalies can be challenging. To address these issues, we have recently described a quantitative technique, allele specific methylated multiplex real-time quantitative PCR (ASMM RTQ-PCR) detecting low degree of LOI, requiring small quantity of DNA and interrogating at least 3 different CpG [23].

Genotype/Phenotype Correlations

RSS patients carrying ICR1 LOM often display a more severe abnormal phenotype regarding growth parameter associated with a more typical dysmorphy (relative macrocephaly, prominent forehead) and highly evocative body asymmetry [12, 19, 20].

Specific features of mUPD7 RSS patients are mild developmental delay mainly consisting in speech difficulties, predisposition to myoclonus dystonia and a putative susceptibility to develop autism traits [20, 24, 25]. All these features are thought to be related to disruption of expression of specific imprinted genes on chromosome 7.

Clinical Guidelines

A multidisciplinary clinical team aware of the particularities of RSS can be very helpful for these patients. In the first years of life, the nutritional support and the gastrointestinal issues (atypical reflux, constipation, lack of appetite) are of major concern. It is important to bear in mind that the nutritional status should be improved before the initiation of GH therapy. However, one should be extremely careful in the nutritional goals to reach to avoid the onset of adrenarche with rapid bone age progression and precocious or early central puberty, insulin resistance and latter metabolic complications. GH therapy is not easy to manage in this group of patients because of frequent elevated IGF-1 serum levels, especially in the RSS group of patients with 11p15 epimutation. However, GH treatment can help not only for the height gain but also improves the muscular strength, diminished the fatigability and might ameliorate food intake. RSS patients can present scoliosis, leg limb discrepancy that will require a specialized orthopedic care but also ear nose and throat, orthodontic, kidney, cardiac, and neurologic, behavior and cognitive work-up and follow-up. Their feeding and language difficulties will often require food and language therapy. Their muscular weakness should be early diagnosed and treated by physical therapy if necessary and physical activity should be encouraged to develop the muscle mass and strength.

Beckwith-Wiedemann Syndrome

Clinical Diagnosis
BWS is a rare overgrowth disorder involving developmental abnormalities, tissue and organ hyperplasia and an increased risk of childhood tumors [26]. It is generally accepted that diagnosis of BWS requires at least 3 clinical findings including at least 2 major signs which are macroglossia, macrosomia at birth or postnatal overgrowth, abdominal wall defect (ranging from umbilical hernia to exomphalos), organomegaly. Minor findings include neonatal hypoglycemia, ear creases and pits, facial nevus, hemihyperplasia, and embryonal tumors.

Molecular Defect
BWS results from molecular or chromosomal alterations that cause overexpression of the paternally expressed genes or a lack of expression of the maternally expressed genes within the 11p15 region [26, 27] (fig. 2):
- loss of methylation (LOM) at ICR2 (50–60% of cases),
- uniparental disomies of paternal origin (pUPD) (20% of cases),
- gain of methylation (GOM) at ICR1 (5–10%),
- genetic mutations into *CDKN1C* gene on the maternal allele (5% of cases),
- cytogenetic anomalies (1–2% of cases) consisting of maternally inherited
  balanced rearrangements (translocations or inversions) and trisomy with double
  dose of the paternal 11p15 region (resulting from duplications or unbalanced
  reciprocal translocation involving the 11p15 region).

  All those molecular defects display a mosaic pattern except *CDKN1C* mutations and cytogenetic anomalies.

Genotype/Phenotype Correlations
Genetic diagnosis and molecular characterization are important to determine the outcome of BWS and perform reliable genetic counseling. Particularly, pUPD and GOM at ICR1 result in a high risk of nephroblastoma while *CDKN1C* mutations can be inherited and cause familial recurrence [26, 27]. *CDKN1C* mutations are strongly associated with exomphalos whereas macrosomia is more frequent in ICR1 GOM. ICR2 loss of methylation is the most frequent molecular anomaly associated with BWS. It leads to a lower tumor risk during childhood than ICRI GOM with no occurrence of nephroblastoma but the possibility of other tumors like hepatoblastoma or neuroblastoma [28, and unpublished data].

Human Multilocus Imprinting Disorders

Imprinting disorders are complex syndromes and display a high degree of clinical heterogeneity. Until a few years ago, it was believed that LOI leading to various human syndromes were isolated events affecting only a given locus involved in a particular syndrome. However, we found that a subset of BWS patients, including some of them born after assisted reproductive technology (ART), displayed LOM at imprinted loci other than the ICR2 11p15 region [28]. Concomitantly, Mackay et al. [29] reported a subset of TNDM patients with 6q24 LOM (*ZAC1* DMR) also displaying multilocus imprinting defects. Since these two reports, other groups have described multilocus imprinting defects in BWS and in TNDM patients. More recently, we revealed that about 10% of RSS patients exhibit multilocus LOM at both paternally and maternally methylated loci and that over two-thirds of these patients exhibit LOM at *DLK1/GTL2* IG-DMR 14q34 in addition to that at ICR1 11p15 [30]. These regions are two of the three paternally methylated ICRs. Comparisons between the clinical characteristics of monolocus LOM patients and multilocus LOM patients did not reveal any statistically significant differences. One of the explanations could be that the dominant phenotype is determined by the locus with the more extensive hypomethylation. However, we observed that some RSS patients have similar degrees of LOM at different loci but have only the RSS phenotype. This led us to suggest an epidominance effect exerted by the 11p15 region in the expression of the phenotype [30, 31]. However, imprinting disorders have complex phenotypes and some of them share some clinical characteristics, for example TNDM and BWS (abdominal defects and macroglossia) and RSS and chromosome 14-related syndromes (intrauterine growth retardation). For these overlapping features, it is difficult to relate the observed clinical abnormality to one or another given locus. Most multilocus defects are partial, thus reflecting a degree of mosaicism. The expression of one syndrome rather than another may possibly therefore be explained by tissue-specific mosaicism.

Assisted Reproductive Technologies and LOI

Several studies reported the increased incidence of Assisted Reproductive Technologies (ART) conception in BWS and Angelman syndrome patients [9, 32]. This finding seems also relevant for RSS patients with ICR1 LOM [19, and unpublished data]. In mice, in vitro fertilization (IVF) and embryo culture lead to loss of imprinting of several genes including *IGF-2* and *H19* [32]. To date, no specific procedure has been identified. It appears that the environment acts through epigenetic mechanisms to modulate development. Imprinting defect in patients born after ART illustrates the link between early conception environment and epigenetic changes. Larger series are needed to investigate this further. Identification of the factors involved in the maintenance and/or the establishment of imprinting is undoubtedly crucial for understanding both the mechanisms underlying imprinting regulation and which disruptions lead to complex diseases such as RSS and BWS.

Identified Molecular Etiologies of RSS and BWS

Genetic defects elements of *cis*-acting at the 11p15 locus have been identified both in BWS and RSS but remain uncommon and mostly identified in BWS patients with ICR1 gain of methylation. However, further studies analyzing large series of BWS or RSS patients with extensive analysis of large DNA regions, not only restricted to the ICR regions, are needed to better characterize the frequency of these *cis*-acting genetic defects leading to epigenetic anomalies.

trans-acting factor anomalies have been involved in the pathogenesis of rare cases of human multilocus imprinting syndromes and may also be involved in some cases of BWS and RSS. Further studies analyzing other *trans*-acting factors in the large series of patients or in informative familial cases are also needed [33–35].

Conclusions

The identification of ICR1 LOM and mUPD7 in nearly 70% of patients with RSS is a considerable step towards understanding the pathophysiology of this disorder. It involves imprinted genes like *IGF-2* and *H19* within the 11p15 region. It also involves imprinted genes on chromosome 7 that remain unidentified. In the same way, it is not established if there is an interaction between chromosome 7-encoded factors and the growth relevant region 11p15. Long-term, systematic endocrine follow-up of cohorts of RSS patients with ICR1 LOM and mUPD7 will help to precise phenotypic specificity and to draw appropriate guidelines for the clinical management. In about 30% of patients with a clinical diagnosis of RSS, the underlying molecular defect is unknown. However, the relatively nonspecific features of RSS present a continuing challenge to clinical diagnosis. This is why the use of a clinical score can help reduce the heterogeneity of patients described in the literature with idiopathic RSS. For BWS, the molecular characterization of the 11p15 defect is crucial for an adequate tumor clinical detection and follow-up. The clinical and molecular overlap between some of the different human imprinting syndromes needs to be better characterized. The existence of multilocus imprinting disorders makes this task even more complex. The precise identification of *cis*-acting and *trans*-acting genetic defects leading to 11p15 epimutation in BWS and RSS is very important for genetic counseling.

Acknowledgments

We thank the patients and their families and the physicians of the affected individuals.

This work was supported by recurrent INSERM funding and ANR EPIFEGRO 2010 (IN, YLB: partner 1). S.A. was supported by the INSERM-ANR EPIFEGRO 2010.

References

1 Gicquel C, El Osta A, Le Bouc Y: Epigenetic regulation and fetal programming. Best Pract Res Clin Endocrinol Metab 2008;22:1–16.

2 Rahman N: Mechanisms predisposing to childhood overgrowth and cancer. Curr Opin Genet Dev 2005; 15:227–233.

3 Lewis A, Reik W: How imprinting centres work. Cytogenet Genome Res 2006;113:81–89.

4 Morgan HD, Santos F, Green K, et al: Epigenetic reprogramming in mammals. Hum Mol Genet 2005;14:R47–R58.

5 Surani MA, Hayashi K, Hajkova P: Genetic and epigenetic regulators of pluripotency. Cell 2007;128: 747–762.

6 Thorvaldsen JL, Fedoriw AM, Nguyen S, et al: Developmental profile of H19 differentially methylated domain (DMD) deletion alleles reveals multiple roles of the DMD in regulating allelic expression and DNA methylation at the imprinted H19/IGF-2 locus. Mol Cell Biol 2006;26:1245–1258.

7 Tang WY, Ho SM: Epigenetic reprogramming and imprinting in origins of disease. Rev Endocr Metab Disord 2007;8:173–182.

8 Jirtle RL, Skinner MK: Environmental epigenomics and disease susceptibility. Nat Rev Genet 2007;8: 253–262.

9 Gicquel C, Gaston V, Mandelbaum J, et al: In vitro fertilization may increase the risk of Beckwith-Wiedemann syndrome related to the abnormal imprinting of the KCN1OT gene. Am J Hum Genet 2003;72:1338–1341.

10 Robertson KD: DNA methylation and human disease. Nat Rev Genet 2005;6:597–610.

11 Gicquel C, Rossignol S, Cabrol S, et al: Epimutation of the telomeric imprinting center region on chromosome 11p15 in Silver-Russell syndrome. Nat Genet 2005;37:1003–1007.

12 Netchine I, Rossignol S, Dufourg MN, et al: 11p15 imprinting center region 1 loss of methylation is a common and specific cause of typical Russell-Silver syndrome: clinical scoring system and epigenetic-phenotypic correlations. J Clin Endocrinol Metab 2007;92:3148–3154.

13 Gaston V, Le Bouc Y, Soupre V, et al: Analysis of the methylation status of the KCNQ1OT and H19 genes in leukocyte DNA for the diagnosis and prognosis of Beckwith-Wiedemann syndrome. Eur J Hum Genet 2001;9:409–418.

14 Silver HK, Kiyasu W, George J, et al: Syndrome of congenital hemihypertrophy, shortness of stature, and elevated urinary gonadotropins. Pediatrics 1953;12:368–376.

15 Russell A: A syndrome of intra-uterine dwarfism recognizable at birth with cranio-facial dysostosis, disproportionately short arms, and other anomalies (5 examples). Proc R Soc Med 1954;47:1040–1044.

16 Price SM, Stanhope R, Garrett C, et al: The spectrum of Silver-Russell syndrome: a clinical and molecular genetic study and new diagnostic criteria. J Med Genet 1999;36:837–842.

17 Bartholdi D, Krajewska-Walasek M, Ounap K, et al: Epigenetic mutations of the imprinted IGF-2-H19 domain in Silver-Russell syndrome (SRS): results from a large cohort of patients with SRS and SRS-like phenotypes. J Med Genet 2009;46:192–197.

18 Bruce S, Hannula-Jouppi K, Peltonen J, et al: Clinically distinct epigenetic subgroups in Silver-Russell syndrome: the degree of H19 hypomethylation associates with phenotype severity and genital and skeletal anomalies. J Clin Endocrinol Metab 2009;94:579–587.

19 Wakeling EL, Amero SA, Alders M, et al: Epigenotype-phenotype correlations in Silver-Russell syndrome. J Med Genet 2010;47:760–768.

20 Binder G, Begemann M, Eggermann T, et al: Silver-Russell syndrome. Best Pract Res Clin Endocrinol Metab 2011;25:153–160.

21 Anderson J, Viskochil D, O'Gorman M, et al: Gastrointestinal complications of Russell-Silver syndrome: a pilot study. Am J Med Genet 2002;113: 15–19.

22 Preece MA: The genetics of the Silver-Russell syndrome. Rev Endocr Metab Disord 2002;3:369–379.

23 Azzi S, Steunou V, Rousseau A, et al: Allele-specific methylated multiplex real-time quantitative PCR (ASMM RTQ-PCR), a powerful method for diagnosing loss of imprinting of the 11p15 region in Russell-Silver and Beckwith-Wiedemann syndromes. Hum Mutat 2011;32:249–258.

24 Guettard E, Portnoi MF, Lohmann-Hedrich K, et al: Myoclonus-dystonia due to maternal uniparental disomy. Arch Neurol 2008;65:1380–1385.

25 A genomewide screen for autism: strong evidence for linkage to chromosomes 2q, 7q, and 16p. Am J Hum Genet 2001;69:570–581.

26 Gicquel C, Rossignol S, Le Bouc Y: Beckwith-Wiedemann syndrome. Orphanet Encyclopedia 2005 (http://www.orpha.net/data/patho/GB/uk-BWS05.pdf).

27 Cooper WN, Luharia A, Evans GA, et al: Molecular subtypes and phenotypic expression of Beckwith-Wiedemann syndrome. Eur J Hum Genet 2005;13: 1025–1032.

28 Rossignol S, Steunou V, Chalas C, et al: The epigenetic imprinting defect of patients with Beckwith-Wiedemann syndrome born after assisted reproductive technology is not restricted to the 11p15 region. J Med Genet 2006;43:902–907.

29 Mackay DJ, Boonen SE, Clayton-Smith J, et al: A maternal hypomethylation syndrome presenting as transient neonatal diabetes mellitus. Hum Genet 2006;120:262–269.

30 Azzi S, Rossignol S, Steunou V, et al: Multilocus methylation analysis in a large cohort of 11p15-related foetal growth disorders (Russell-Silver and Beckwith-Wiedemann syndromes) reveals simultaneous loss of methylation at paternal and maternal imprinted loci. Hum Mol Genet 2009;18:4724–4733.

31 Azzi S, Rossignol S, Le Bouc Y, et al: Lessons from imprinted multilocus loss of methylation in human syndromes: a step toward understanding the mechanisms underlying these complex diseases. Epigenetics 2010;5:373–377.

32 Amor DJ, Halliday J: A review of known imprinting syndromes and their association with assisted reproduction technologies. Hum Reprod 2008;23:2826–2834.

33 Demars J, Gicquel C: Epigenetic and genetic disturbance of the imprinted 11p15 region in Beckwith-Wiedemann and Silver-Russell syndromes. Clin Genet 2012;81:350–361.

34 Demars J, Rossignol S, Netchine I, et al: New insights into the pathogenesis of Beckwith-Wiedemann and Silver-Russell syndromes: contribution of small copy number variations to 11p15 imprinting defects. Hum Mutat 2011;32:1171–1182.

35 Demars J, Shmela ME, Rossignol S, et al: Analysis of the IGF-2/H19 imprinting control region uncovers new genetic defects, including mutations of OCT-binding sequences, in patients with 11p15 fetal growth disorders. Hum Mol Genet 2010;19:803–814.

Irène Netchine, MD, PhD
APHP, Hôpital Armand-Trousseau, Explorations Fonctionnelles Endocriniennes
UPMC, INSERM U938 team 4
26, avenue du Docteur Arnold-Netter, FR–75012 Paris (France)
Tel. +33 1 44 73 64 48, E-Mail irene.netchine@trs.aphp.fr

Mullis P-E (ed): Developmental Biology of GH Secretion, Growth and Treatment.
Endocr Dev. Basel, Karger, 2012, vol 23, pp 71–85 (DOI: 10.1159/000341755)

Bioinformatics Tools and Databases for the Study of Human Growth Hormone

Amit V. Pandey

Pediatric Endocrinology, Diabetology and Metabolism, Department of Clinical Research, University of Bern, and
University Children's Hospital Bern, Bern, Switzerland

Abstract

Advances in novel molecular biological diagnostic methods are changing the way of diagnosis and
study of metabolic disorders like growth hormone deficiency. Faster sequencing and genotyping
methods require strong bioinformatics tools to make sense of the vast amount of data generated
by modern laboratories. Advances in genome sequencing and computational power to analyze the
whole genome sequences will guide the diagnostics of future. In this chapter, an overview of some
basic bioinformatics resources that are needed to study metabolic disorders are reviewed and some
examples of bioinformatics analysis of human growth hormone gene, protein and structure are
provided.

Advances in molecular genetics and bioinformatics are changing the ways of detection and treatment of genetic diseases. Novel DNA sequencing and mapping tools together with the availability of the human genome have made it possible to identify the molecular basis of complex genetic disorders. Study of human growth hormone (GH) has been at the forefront of advances in modern diagnostic and treatment methods. In the 21st century, endocrinology has evolved from the measurement of hormones and metabolites to genetic analysis of whole pathways involved in hormone synthesis and action, including biosynthesis of hormones, genes and proteins involved in hormone synthesis, interaction of hormones with their receptors as well as regulation of genes involved in hormone biosynthesis and action pathways in a cellular and tissue-specific manner. The importance of human GH in clinical medicine has been recognized for a long time. Human GH was one of the first genes to be studied at the molecular genetic level and pioneered the molecular biological studies for cloning in bacteria and production as recombinant molecules [1, 2]. Human GH was also among the first recombinant proteins that was made for therapeutic use and contributed to the establishment of the pioneering biotechnology company Genentech in San Francisco, Calif., USA.

Molecular biological methods allow detection of rare genetic disorders when biochemical laboratory measurements are not clear as well as in pediatric and neonatal diagnosis where sample size is often limited. The latest advancements in the field have been the availability of the human genome sequence and novel sequencing technologies. An almost complete catalogue of human genes has allowed endocrinologists to identify candidate genes for many genetic disorders and subsequent sequencing of genetic material from patients had led to identification of the molecular basis of many hormonal and metabolic defects.

Basic Analytical Methods

Identification of the exact location of mutations is of importance to clinical laboratories as a wide variation in severity of disease could be seen in disorders related to GH, often due to the wide nature of point mutations in the GH1 gene with each mutation resulting in a specific effect. Functional genomics characterization is performed by identification of the exact location of mutations followed by the biochemical characterization of recombinant genes/proteins that replicate the patient DNA sequences in an experimental setup. Modern diagnostic techniques employ DNA amplification-based methods and require only a small amount of DNA for most diagnostic methods.

PCR and Sequencing
It is advisable to use primer design software tools available at bioinformatics resources rather than relying on manual design. Primer3Plus provides a web interface to the popular Primer3 program with a task oriented approach. Several different options for primer design are available including sequencing and cloning as well as design of probes for Northern blot analysis, etc. (http://www.bioinformatics.nl/cgi-bin/ primer3plus/primer3plus.cgi). Primers that display 'significant' homology to alternative targets should be discarded. The primer blast tool available at NCBI can screen for mismatches and alternative binding sites (http://www.ncbi.nlm.nih.gov/tools/ primer-blast/). DNA sequencing is still the most reliable method to confirm a genetic defect and provide the exact location and nature of the disorder. Sequencing of DNA is generally no longer performed in individual laboratories and has been relegated to central facilities or outsourced to private companies providing sequencing services. The automated and efficient sequencing technology has allowed comparison of normal and patient DNA sequences without manually reading the chromatograms. One efficient tool for automated mutation analysis from DNA sequence traces is Mutation Surveyor software from Softgenetics (http://www.softgenetics.com/mutationSurveyor.html). Mutation Surveyor is available for academic labs with simultaneous reading of 48 samples from Sanger Sequencing traces provided by automated dye terminator sequencers. The Mutation Surveyor can locate genetic variants, SNPs and

Indel mutations between reference traces and sample/patient traces and has a several reporting possibilities for both research and laboratory diagnostics.

Diagnostics of the Future: SNP/Mutation Genotyping and Next Generation Sequencing
The high-density oligonucleotide SNP arrays are formed by spotting a large number of probes on small chips, allowing multiple SNPs to be probed simultaneously [3]. Oligonucleotide microarray methods currently have lower specificity and sensitivity compared to sequencing but the amount of SNPs that can be checked in a single analysis is of great advantage. Advances in a new generation of SNP chips that can include known disease-causing mutations will make it possible to screen for variations in GH1 and related genes in a single chip based assay. In the longer term, a whole genome sequencing approach may replace current single gene detection approaches. In the past few years, the cost of genome sequencing has come down considerably. Once the costs drop below USD 1,000 per genome and reliability can be of clinical diagnostic standards, it would be possible to order whole genome sequencing of newborns rather than multiple individual genetic tests. Such an approach would likely be cost effective when multiple blood drawings, individual sequencing and analysis that is currently used is replaced by a single genome sequencing test. This will allow analysis of not only GH1 but all related genes and pathways, promoters, inducers, transcription factors etc. that are likely to have an impact on GH1 based effects. Moreover, once the whole genome sequence of a patient is available, different queries can be performed to check for any individual gene that will allow clinicians to get the answers in a very fast manner compared to currently available methods.

Functional Analysis with Recombinant Proteins and Cell-Based Assays
Once a defect has been identified, the GH gene from patients can be cloned directly, or often the mutation/disordered can be replicated by changing the normal gene that is already available in laboratories in the form of standard DNA libraries. It is then possible to deduce the amino acid sequence of proteins and predict its structural and biophysical properties. Recombinant genes with artificially introduced mutations/defects are introduced in bacteria/yeast/mammalian cells to generate proteins, which are then studied to identify enzymatic or biochemical disruptions to gain information about possible causes of disease [4]. In cases where structural information of proteins is available, an in silico modeling approach could be used to predict the effects of mutations [5, 6].

Molecular Genetics of Human Growth Hormone

The GH gene (NG_011676.1, gene id: 2688, OMIM 139250) is located on chromosome 17 and contains 5 exons separated by 4 introns. There are several isoforms of GH mRNA due to alternative splicing that produce proteins of different sizes.

1. NM_000515.3, NP_000506.2 GH1 has all five exons, generating the predominant 22-kDa isoform 1.

2. NM_022559.2 NP_072053.1 GH2 variant utilizes an alternative splice acceptor site 45 nt into exon 3 to generate the 20-kDa isoform 2.

3. NM_022560.2 NP_072054.1 GH3 variant is missing exon 3.

4. NM_022561.2 NP_072055.1 GH4 variant is missing exons 3 and 4.

5. NM_022562.2 NP_072056.1 GH5 variant is missing exons 2, 3 and 4, leading to a frameshift that generates the shortest isoform 5. The majority of the sequence encoding the signal peptide is missing, including the signal cleavage site. The carboxy-terminus of isoform 5 is unique among all other GH1 isoforms.

Human GH binds two molecules of GHR (OMIM 600946) that leads to receptor dimerization and starts a signaling cascade of downstream effects. The regulation of GH synthesis and secretion is governed by a number of genes that include the transcription factors PROP1 (OMIM 601538) and PIT1 (OMIM 173110). Release of GH is dependent on GHRH (OMIM 139190) and GHRHR (OMIM 139191) genes. The GHRH is synthesized and released from the hypothalamus, and translocates to the pituitary to bind to GHRHR which sends a signal into the somatotroph that release the GH stored in secretory granules. The binding of GH to GHR molecules generates a signal to produce IGF1 (OMIM 147440). The GH molecules that are bound to membrane-anchored GH receptors can be released into the circulation by excision of the extracellular portion of the GHR molecules. In one of the earliest molecular genetic studies Philips et al. [7] found that the GH gene was deleted in 2 families with type IA GH deficiency but the GH genes of persons with type IB seemed normal. In a larger study, Mullis et al. [8] analyzed GH1 DNA from 78 subjects with severe IGHD. Among the 78 patients 10 had GH1 deletion; 8 had a 6.7-kb deletion and 2 had a 7.6-kb deletion. Isolated GH deficiency has been divided into four different forms.

IGHD Type 1A

IGHD 1A is the classical form of GH deficiency and results from mutations in the GH1 gene. In the IGHD 1A, deletions, frameshifts, and nonsense mutations lead to loss of GH and cause severe dwarfism.

IGHD-1B

Isolated GH deficiency type IB can be caused by mutation in either the GH1 (OMIM 139250) or GHRHR (OMIM 139191) gene. Patients with IGHD type IB may have low but detectable levels of GH, short stature, significantly retarded bone age, and a positive response to GH replacement therapy.

IGHD Type 2

The IGHD type 2 results from aberrant splicing of the GH1 gene. In affected members of a Turkish family segregating IGHD II, Phillips and Cogan [9] identified a splice site mutation in the GH1 gene. In 2005, Mullis et al. [10] studied 57 subjects with IGHD-2 with different splice site as well as missense mutations in the GH1 gene. The subjects with the splice site mutation causing the skipping of exon 3 were more likely to develop other pituitary hormone deficiencies.

IGHD Type 3
The IGHD type 3 is a compound phenotype due to mutation in the BTK gene (OMIM 300300), that is mutated in X-linked agammaglobulinemia [11].

Kowarski Syndrome
Kowarski et al. [12] studied 2 boys with growth retardation and delayed bone ages and suggested that the phenotype may be caused by the secretion of a biologically inactive but immunoreactive GH IGHD. Takahashi et al. [13] reported a boy with short stature and heterozygosity for a mutant GH gene. In this child, the GH not only could not activate the GHR but also inhibited the action of wild-type GH because of its greater affinity for GHR creating a dominant-negative effect. Besson et al. [14] described a Serbian patient with Kowarski syndrome who was homozygous for a mutation in the GH1 gene disrupting the first disulfide bridge in a GH, while the heterozygous parents were of normal stature.

Bioinformatics Methods and Databases for Disease Information

There are three major portals for accessing bioinformatics information that can be used for analysis of GH and related genes. The information provided in following pages can be applied to any gene or protein: in Europe (European Bioinformatics Institute (EBI) http://www.ebi.ac.uk), the United States (National Center for Biotechnology Information (NCBI), http://www.ncbi.nlm.nih.gov) or the National Institute of Genetics (NIG) in Japan (http://www.nig.ac.jp). These three databases are the main portals for all genetic sequence information and are synchronized every day to reflect the same information across all three repositories. In addition, UniProt/ SwissProt (www.uniprot.org) is all about proteins and contains a vast array of direct links to other databases and tools for the protein entries. The NCBI is one of the main portals of biological data and serves as a central resource. There are multiple tools and databases available through the NCBI portal. While performing searches it is important to find the reference sequence numbers and then use LinkOut tools located on the right hand side of the browser window to go to other related tools and databases related to gene/protein rather than performing new searches. There are multiple entries for same genes and proteins in the Genbank databases, and LinkOut service ensures that correct information specific for a particular gene or protein variant is reached by following the links. A list of database accession numbers for the human growth hormone is given in table 1. When searching for disease or gene-related information these accession numbers should be used rather than text base search.

Table 1. Common database accession numbers for human GH

Database resource	Accession no.
RefSeq protein	NP_000506
RefSeqRNA	NM_000515
UniGene	Hs.655229
Uniprot	P01241
GeneID	2688
GeneCards	GC17M061994
GHNC	HGNC:4261
PANTHER	PTHR11417
PDB	1A22, 1AXI, 1BP3, 1HGU, 1HUW, 1HWG, 1HWH, 1KF9, 3HHR

Entrez Gene

The NCBI gene database (www.ncbi.nlm.nih.gov/gene) provides gene-specific information (map, sequence, expression, structure, function, citation, and homology data). Unique identifiers are assigned to genes with defining sequences, genes with known map positions, and genes inferred from phenotypic information. These gene identifiers are used throughout NCBI's databases and tracked through updates of annotation.

OMIM

The Online Mendelian Inheritance in Man is a database of diseases and genes. It was initiated by Dr. Victor A. McKusick, and published in printed form for 12 editions and later been adopted to online version. Every disease and gene is assigned a six-digit number with first digit being related to method of inheritance: 1 for autosomal-dominant, 2 for autosomal-recessive, 3 for X-linked phenotype, 4 for Y-linked phenotype, 5 for mitochondrial and 6 for autosomal phenotype catalogued after 1994 for both autosomal-dominant and autosomal-recessive phenotypes. An asterix represents an entry with a known sequence, a number symbol (#) indicates a descriptive phenotype entry, a plus sign indicates a gene of known sequence and phenotype, and a percentage (%) sign indicates that the entry has a confirmed mendelian phenotype where the molecular bases of disease and gene are not known. The OMIM database entries for GH1 and related genetic disorders are shown in table 2.

Basic Alignment Search Tool (BLAST)

The BLAST is a powerful technique to analyze DNA or protein sequences by comparing a query sequence against whole databases (http://www.ncbi.nlm.nih.gov/blast). The BLAST can perform several different searches including nucleotide to nucleotide search (Blast N), protein to protein searches (Blast P) or searching a nucleotide entry against protein sequence databases (Blast X). After a sequence from patient DNA has been obtained, BLAST is the first tool that a researcher uses to compare that sequence

Table 2. OMIM accession numbers for GH1 and related genetic disorders

Disease	Gene	OMIM no.
Combined pituitary hormone deficiency	PIT-1	#613038
GH, TSH, Prl	PIT-1 (POU1F1)	
GH, TSH, Prl and LH/FSH	PROP-1	#262600
GH, TSH, Prl and LH/FSH ACTH	HESX-1	
With/without midline defects		
GH, TSH, Prl and LH/FSH ACTH	LHX-3	#221750
With cervical spine shortening		
GH, TSH, Prl and LH/FSH ACTH	LHX-4	#262700
Growth hormone	GH1	*139250
Isolated growth hormone deficiency type IA	GH1	#262400
Isolated GH deficiency type II	GH1	#173100
Isolated GH deficiency type IB	GH1	#612781
Isolated GH deficiency type IB	GHRHR	*139191
Kowarski syndrome	GH1	#262650
Isolated GH deficiency type III	BTK	#307200
GH resistance, high GH low IGF-1	GHR	#262500

against standard sequences. A specialized version of BLAST called align two sequences (bl2seq) can be used to align patient DNA again standard sequence from databases for both nucleotide or translated protein sequences to identify differences.

RefSeq

Primary DNA sequences are submitted by individual researchers or sequencing projects and are assigned an accession number. These primary sequences are often redundant and contain several copies of the same sequence submitted by different laboratories. Therefore, a separate RefSeq project (http://www.ncbi.nlm.nih.gov/RefSeq/) was established to provide a reference sequence for each molecule in a nonredundant dataset that is verified, checked for accuracy and catalogued as reference sequences, where each gene or protein has only one entry. The RefSeq entries are distinguished from other entries by a distinct accession number scheme. The refseq entries follow a 2 + 6 numbering system, a two-letter code indicating the type of reference sequence followed by an underscore and a six-digit number. The reference sequence entries contain additional information about the gene/protein like major publications, corresponding biological process, gene structure, etc.

Gene Expression Omnibus (GEO)

The GEO portal at NCBI (http://www.ncbi.nlm.nih.gov/geo/) is a public repository of gene expression and functional genomics data. Multiple tools are available to analyze and download datasets.

Table 3. List of nonsynonymous variants in human GH1
(NP_000506)

Top of form

SNP ID	Original AA	AA position	Variant AA
rs151263636	ALA	13	VAL
rs71640273	ARG	16	CYS
rs140787052	ALA	17	THR
rs71640274	ARG	19	HIS
rs145327622	HIS	21	TYR
rs61735357	PHE	25	TYR
rs71640276	ASN	47	ASP
rs137853222	CYS	53	PHE
rs137853222	CYS	53	SER
rs61762497	GLU	56	ASP
rs137853220	ARG	77	CYS
rs6174	SER	79	CYS
rs150761983	GLN	84	GLU
rs141217174	SER	85	PRO
rs142396035	GLN	91	ARG
rs137882374	SER	95	ARG
rs5388	VAL	110	ILE
rs137853221	ASP	112	GLY
rs41295253	PRO	133	HIS
rs41295255	ARG	134	LEU
rs138880244	PHE	139	LEU
rs4080076	ASN	149	LYS
rs151243538	ASP	153	HIS
rs184640372	LEU	157	PHE
rs77096658	ARG	178	CYS
rs41295257	ILE	179	MET
rs148474991	ILE	179	MET
rs137853223	ARG	183	HIS

Bottom of form

Homologene

The homologene database (http://www.ncbi.nlm.nih.gov/sites/entrez?db = homologene) allows automated detection of homologs among the curated genomes.

SNP

The SNP database (http://www.ncbi.nlm.nih.gov/sites/entrez?db = snp) is the repository of single nucleotide polymorphisms and short deletions and insertions. A list of nonsynonymous variants in the human GH1 (NP_000506) gene is shown in table 3.

Pubchem BioAssay

The PubChem BioAssay database (http://www.ncbi.nlm.nih.gov/sites/entrez?db = pcassay) contains bioactivity screens of chemicals. Searches can be performed by descriptions, experimental conditions and specific readouts.

UniProt

The UniProt database (http://www.uniprot.org/) provides links for the functional information on proteins with accurate and updated entries. The core data for each UniProt entry includes amino acid sequence, protein and gene names, description, biological function, and links to RefSeq data and references. There are two categories of entries in the UniProt database, the gold star entries with manually curated information and the silver star entries with computationally generated, unreviewed links. The UniProt entry for human GH (P01241) is manually curated, reviewed and constantly updated with new information. Numbering of amino acids on UniProt is according to full-length protein, while in practice the N-26 form of secreted GH is used for numbering. This should be kept in mind when searching for mutations in UniProt or other databases.

Domain Mapping of Disease Mutations (DMDM)

DMDM (http://bioinf.umbc.edu/dmdm/) database shows disease mutations by its gene, protein or domain location [15]. DMDM can show a domain-level view of human amino acid mutations mapped on the protein (fig. 1). The disease causing mutations and polymorphisms are shown along with functional information (e.g. the binding and catalytic activity of the site and the conservation of that domain location).

University of California Santa Cruz Genome Browser

The UCSC genome browser (http://genome.ucsc.edu) is probably the most useful site for complete genome assemblies and human genetic makeup informatics. The genome browsers can zoom and scroll over single chromosomes to locate and show individual genes with links to expression pattern, homology to other species and related genes. The BLAT tool (not to be confused with BLAST) provides precise information about gene structure and intron-exon boundaries by mapping a DNA sequence over a genome.

Gene Expression Atlas

The Gene Expression Atlas (http://www.ebi.ac.uk/gxa/) is a database of meta-analysis summary statistics of a curated subset of the ArrayExpress Archive. It allows queries for condition-specific gene expression patterns and searches for biologically interesting samples. Gene expression studies from published data are analyzed and curated, and before performing a separate experiment a search through the gene expression atlas can reveal already available information from previous gene expression experiments. More than 250 experiments related to differentially expressed GH genes are currently archived in the gene expression atlas with links and analysis for organ, disease, cell lines, compound treatment and RNAi experiments.

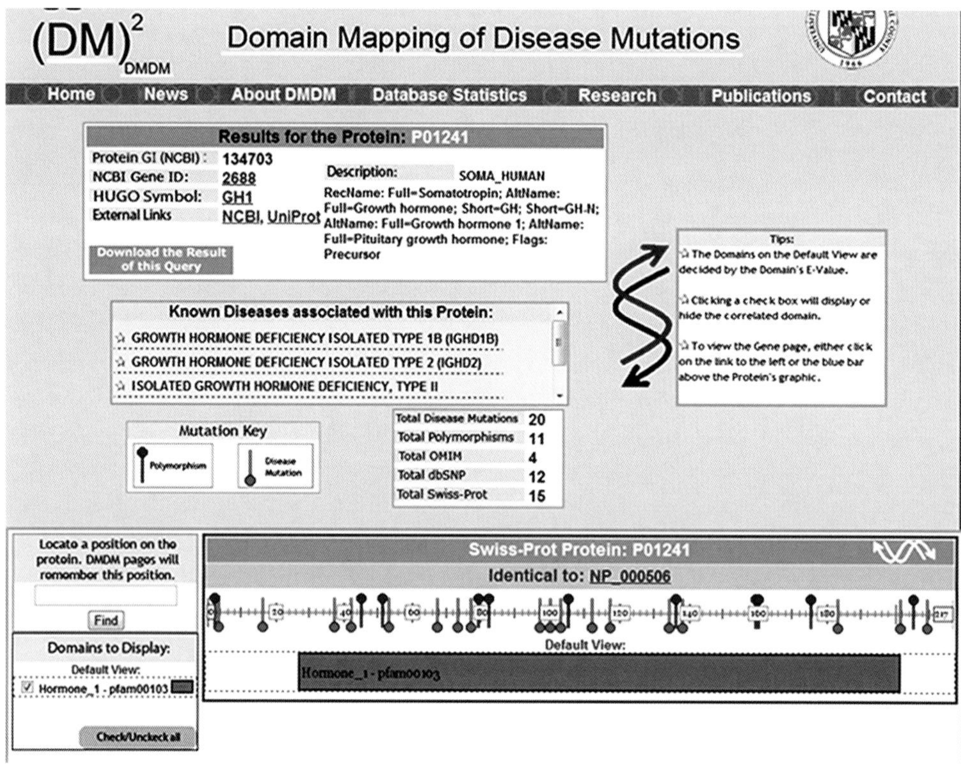

Fig. 1. The DMDM gateway page for human GH showing mapping of mutants and polymorphisms in GH1 gene on GH protein.

GeneReviews and Genetest

GeneReviews are expert-authored, peer-reviewed, current disease descriptions that apply genetic testing to the diagnosis, management and genetic counseling of patients and families with specific inherited conditions. The reviews are authored and reviewed by known experts in the diagnosis of specific diseases. The articles are peer reviewed and updated by the authors every 2–3 years. Significant changes to clinically relevant information are updated on a more frequent basis. GeneReviews can be searched by disease name, gene symbol, title or author names, etc. Reviews are linked to related literature in PubMed. Genetests is a publicly funded resource created for cataloguing of medical genetics laboratories.

GeneCards and GeneLoc

GeneCards (http://www.genecards.org) is a searchable, integrated database of human genes that provides genetic and functional information on all known human genes. Information catalogued under a GeneCard includes gene name aliases, orthologies, disease relationships, mutations and SNPs, RNA expression data, gene function, pathways, protein, protein-protein interactions, related drugs and compounds, and direct

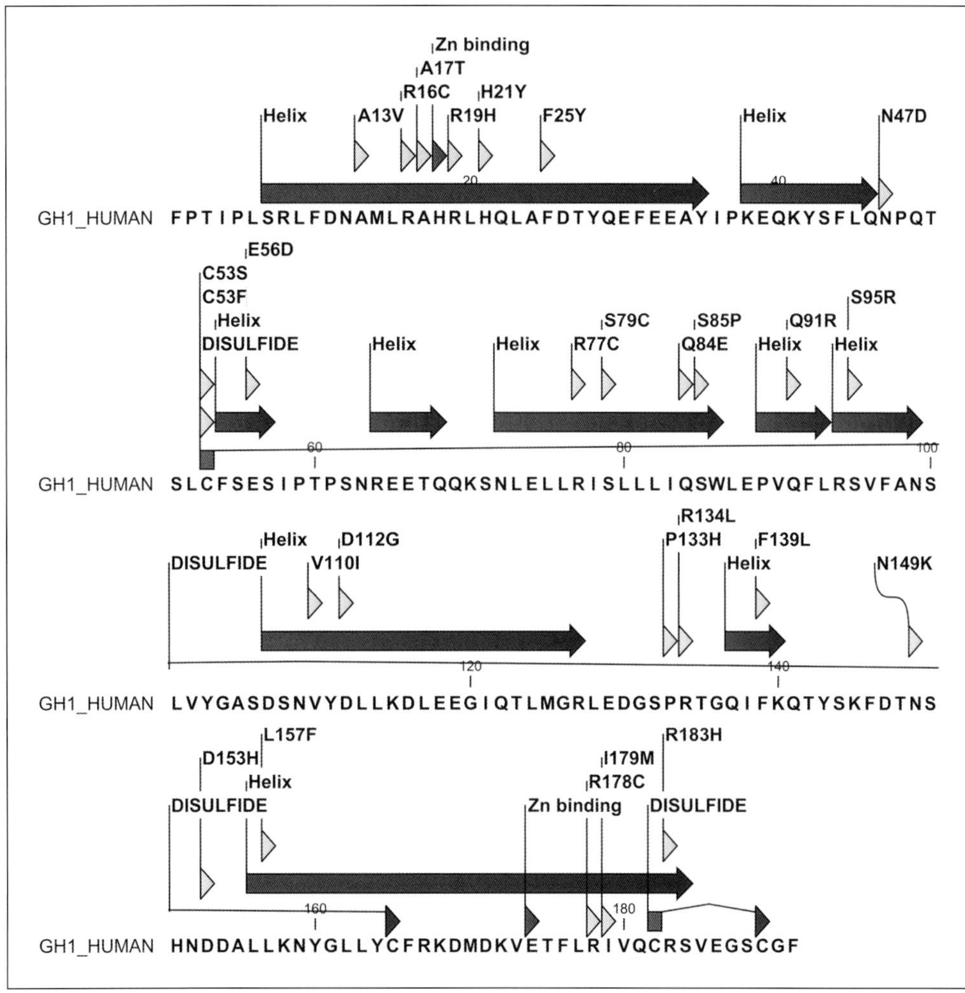

Fig. 2. An overview of the secreted human GH1 (N-26) amino acid sequence. Location of some common mutants and polymorphisms and structural features in GH are shown.

links to suppliers of research reagents and tools such as primers, antibodies, proteins, clones, expression assays and RNAi reagents. GeneLoc records DNA sequence, physical map positions and associated sequences.

Structural Computational Biology Approaches

Structural basis of many GH mutations can be studied by using in silico mutagenesis and molecular dynamic simulations. Detailed primary and secondary structural analysis is performed as a first step (fig. 2, 3). A multiple alignment and sequence conservation analysis can reveal the importance of functional residues (fig. 2). Three-dimensional

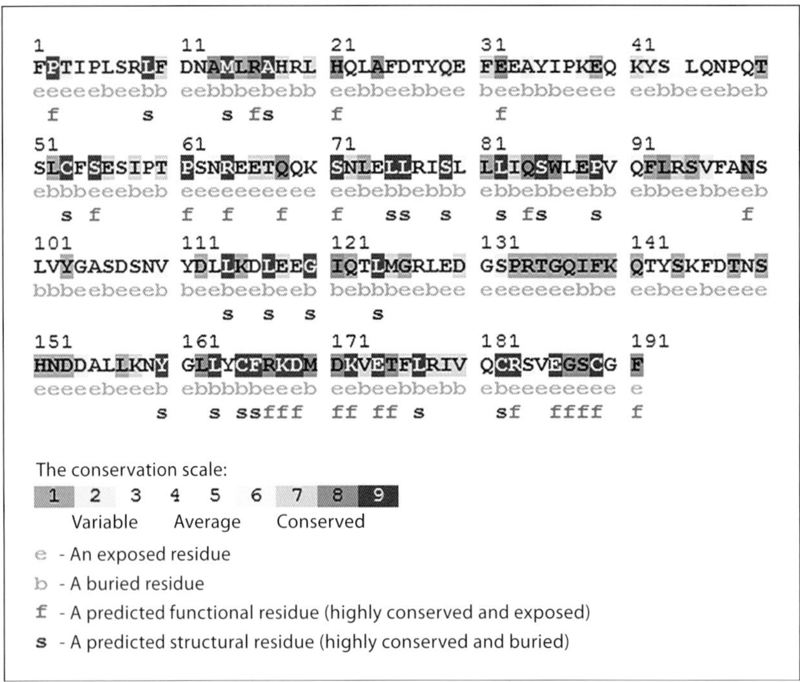

The conservation scale:

Fig. 3. GH1 sequence conservation analysis. e = Exposed residue; b = buried residue; f = predicted functional residue (highly conserved and exposed); s = predicted structural residue (highly conserved and buried).

structures of GH alone and in complex with the extracellular domain of its receptor are known (fig. 4). These structures are used to recreate the mutant molecules of GH and study the properties that are affected by mutation [16, 17]. It is possible to calculate the potential amino acid residues involved in the interactions and predict the effect of mutations at these sites (table 4). Several different aspects are considered that include changes in charge, surface area, location at the binding sites for the receptor, Zn binding sites and dimerization and protein folding. Natural effects of the mutations are studied by molecular dynamic simulations where the GH or GH-GHR complex is put inside a simulator cell filled with water and buffer salts, virtual temperature and pressure controllers are attached to control the environment and changes in properties of mutated proteins are monitored in real time. Analysis of MD trajectories allows following up changes in properties of protein caused by mutations. The 3D structural model of the hGH (NP_000506.2, P01241) sequence (AA 1–191) is built using the human GH isoform 1 structure (1HGU) as a template. Model building is performed with programs YASARA [18] and WHATIF [19]. A secondary structure prediction is used to build the missing parts in the crystal structure using the program DSC [20]. Side chains in newly built parts are then optimized by molecular dynamic (MD) simulations. The MD simulations are performed using the AMBER03 force field [18]. The simulation cell is filled with water, pH is fixed to 7.4 and the AMBER03 [21] electrostatic potentials are evaluated for

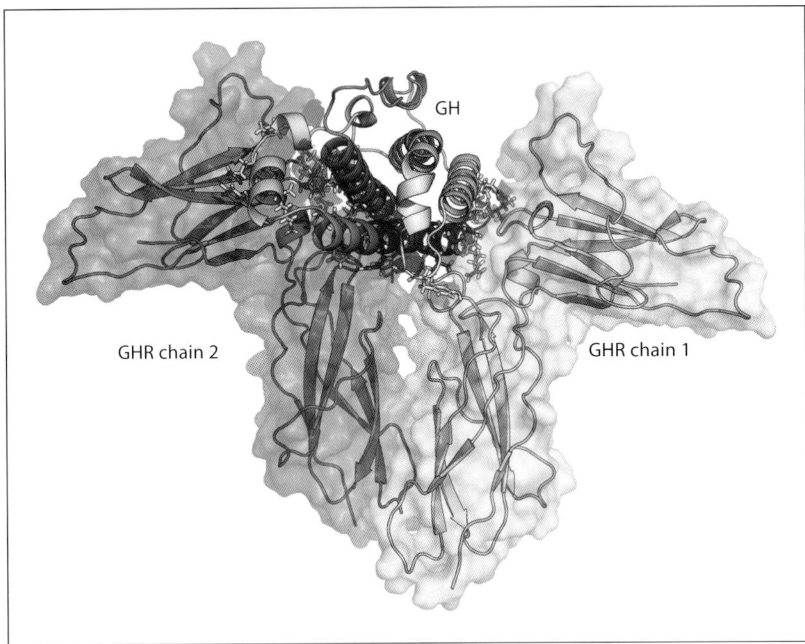

Fig. 4. Complex of GH1 with two molecules of GHR. Amino acids involved in interaction of GH with GHR are shown as stick models, while GH and GHR are shown as ribbons.

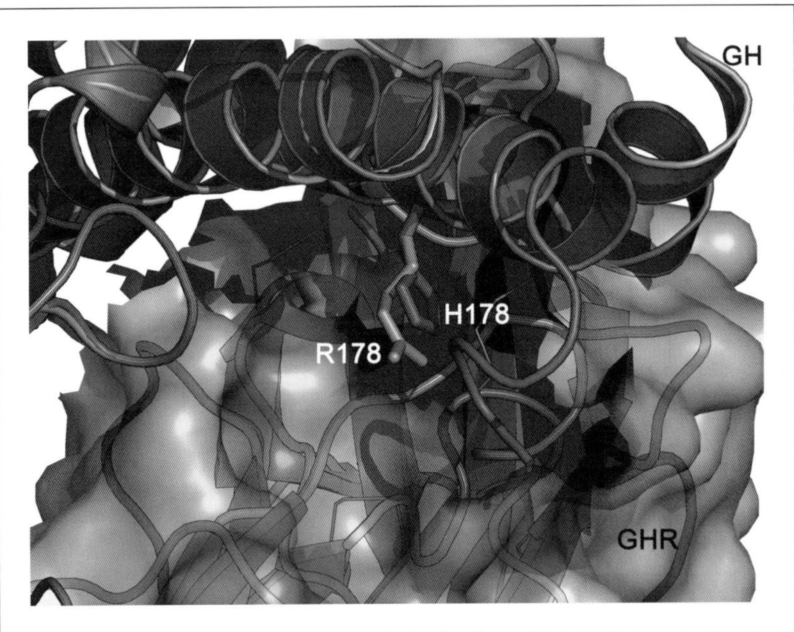

Fig. 5. Example of structural analysis of GH mutations. The analysis of R178H mutations is shown with superimposed models of WT and R178H GH. The side chains of WT R178 and mutant H178 are shown as sticks and the GHR molecule is shown as a surface model. A change from arginine to histidine affects the interaction of GH with GHR leading to reduced binding efficacy.

Table 4. Amino acids involved in formation of the GH-GHR complex

Molecule	Amino acids involved in interaction between GH1 and GHR
GH1	1, 2, 4, 8, 9, 12, 15, 16, 18, 21, 22, 25, 41, 42, 45, 46, 48, 61, 62, 63, 64, 67, 103, 116, 119, 120, 123, 164, 167, 168, 171, 172, 174, 175, 178, 179, 182, 189
GHR chain 1	43, 44, 76, 77, 103, 104, 105, 106, 108, 120, 121, 122, 123, 127, 164, 165, 166, 167, 168, 169, 170, 217, 218, 219, 220
GHR chain 2	43, 44, 71, 102, 103, 104, 106, 122, 123, 126, 165, 168, 169, 216, 219

One molecule of GH forms a complex with two units of GHR. Amino acid numbers correspond to cleaved 191 AA GH. Mutations identified in future that are involved in binding may be predicted to cause a reduced interaction between GH and GHR.

water molecules in the simulation cell and adjusted by the addition of sodium and chloride ions. The MD simulations are run at 298K, 0.9% NaCl and pH 7.4 for 1,000 ps to refine the model. The best models as selected for the analysis and evaluation of mutant amino acids. Structures are checked with the programs WHATCHECK [22], WHATIF [19], Verify3D [23, 24], ERRAT [25] and Ramachandran plot analysis [26, 27]. In silico mutagenesis to create mutants is performed with YASARA and WHATIF.

In silico structural approaches are important to understand the functional implications of individual mutations as determining the structures of each mutation is not feasible. However, exponential advances in computational power have allowed us to study single amino acid changes in GH protein with MD simulations and other bioinformatics approaches described in this chapter. Similar approaches can be applied to other related genes like the GHR, GHRH, etc. Advances in computational power and novel sequencing technologies will bring a different way of disease diagnostics and analysis in the next few years, and bioinformatics skills to analyze the vast amount of sequencing and genomic information will be crucial in the diagnostics of future generations.

Acknowledgements

This work was supported by grants to AVP from the Swiss National Science Foundation (31003A-134926), Bern University Research Foundation and the Schweizerischen Mobiliar Genossenschaft Jubiläumsstiftung.

References

1 Goeddel DV, et al: Direct expression in *Escherichia coli* of a DNA sequence coding for human growth hormone. Nature 1979;281:544–548.

2 Itakura K, et al: Expression in *Escherichia coli* of a chemically synthesized gene for the hormone somatostatin. Science 1977;198:1056–1063.

3 Hoheisel JD: Microarray technology: beyond transcript profiling and genotype analysis. Nat Rev Genet 2006;7:200–210.

4 Flück CE, et al: Mutant P450 oxidoreductase causes disordered steroidogenesis with and without Antley-Bixler syndrome. Nat Genet 2004;36:228–230.

5 Janner M, et al: Clinical and biochemical description of a novel CYP21A2 gene mutation 962_963insA using a new 3D model for the P450c21 protein. Eur J Endocrinol 2006;155:143–151.

6 Petkovic V, et al: Growth hormone (GH) deficiency type II: a novel GH-1 gene mutation (GH-R178H) affecting secretion and action. J Clin Endocrinol Metab 2010;95:731–739.

7 Phillips JA III, et al: Heterogeneity in the molecular basis of familial growth hormone deficiency (IGHD). Am J Hum Genet 1981;33:52A.

8 Mullis PE, et al: Prevalence of human growth hormone-1 gene deletions among patients with isolated growth hormone deficiency from different populations. Pediatr Res 1992;31:532–534.

9 Phillips JA III, Cogan JD: Genetic basis of endocrine disease 6: molecular basis of familial human growth hormone deficiency. J Clin Endocr 1994;78:11–16.

10 Mullis PE, et al: Isolated autosomal dominant growth hormone deficiency: an evolving pituitary deficit? A multicenter follow-up study. J Clin Endocr Metab 2005;90:2089–2096.

11 Duriez B, et al: An exon-skipping mutation in the btk gene of a patient with X-linked agammaglobulinemia and isolated growth hormone deficiency. FEBS Lett 1994;346:165–170.

12 Kowarski AA, et al: Growth failure with normal serum RIA-GH and low somatomedin activity: somatomedin restoration and growth acceleration after exogenous GH. J Clin Endocr 1978;47:461–464.

13 Takahashi Y, et al: Short stature caused by a mutant growth hormone. N Engl J Med 1996;334:432–436.

14 Besson A, et al: Short stature caused by a biologically inactive mutant growth hormone (GH-C53S). J Clin Endocr Metab 2005;90:2493–2499.

15 Peterson TA, et al: DMDM: domain mapping of disease mutations. Bioinformatics 2010;26:2458–2459.

16 Petkovic V, et al: Growth hormone (GH) deficiency type II: a novel GH-1 gene mutation (GH-R178H) affecting secretion and action. J Clin Endocrinol Metab 2010;95:731–739.

17 Petkovic V, et al: A novel GH-1 gene mutation (GH-P59L) causes partial GH deficiency type II combined with bioinactive GH syndrome. Growth Horm IGF Res 2011;21:160–166.

18 Krieger E, et al: Making optimal use of empirical energy functions: force-field parameterization in crystal space. Proteins 2004;57:678–83.

19 Vriend G: WHAT IF: a molecular modeling and drug design program. J Mol Graph 1990;8:52–56.

20 King RD, et al: *DSC*: public domain protein secondary structure predication. Comput Appl Biosci 1997;13:473–474.

21 Liu H, et al: Quantum mechanics simulation of protein dynamics on long timescale. Proteins 2001;44:484–489.

22 Hooft RW, et al: Errors in protein structures. Nature 1996;381:272.

23 Bowie JU, Luthy R, Eisenberg D: A method to identify protein sequences that fold into a known three-dimensional structure. Science 1991;253:164–170.

24 Luthy R, Bowie JU, Eisenberg D: Assessment of protein models with three-dimensional profiles. Nature 1992;356:83–85.

25 Colovos C, Yeates TO: Verification of protein structures: patterns of nonbonded atomic interactions. Protein Sci 1993;2:1511–1519.

26 Ramachandran GN, Ramakrishnan C, Sasisekharan V: Stereochemistry of polypeptide chain configurations. J Mol Biol 1963;7:95–99.

27 Hooft RW, Sander C, Vriend G: Objectively judging the quality of a protein structure from a Ramachandran plot. Comput Appl Biosci 1997;13:425–430.

PD Dr. Amit V. Pandey
Pediatric Endocrinology, Diabetology and Metabolism
Department of Clinical Research, University of Bern
Tiefenaustrasse 120c
CH–3004 Bern (Switzerland)
Tel. +31 308 8044, E-Mail amit@pandeylab.org

Mullis P-E (ed): Developmental Biology of GH Secretion, Growth and Treatment.
Endocr Dev. Basel, Karger, 2012, vol 23, pp 86–95 (DOI: 10.1159/000341761)

Growth Hormone and Cell Growth

Michael J. Waters · Andrew J. Brooks

Institute for Molecular Bioscience, The University of Queensland, St. Lucia, Qld., Australia

Abstract

Growth hormone (GH) promotes stem cell activation, cell proliferation, differentiation and survival, either directly or through the induction of IGF-1. GH acts via its cell membrane receptor to initiate a range of signalling pathways, with JAK2 kinase activation of STAT5 being the most important. The transcription factor STAT5 acts to induce expression of the key growth mediator, IGF-1, but also regulates the expression of a host of other genes, some of which are important growth regulators. In addition to its signalling from the cell membrane, the GH receptor translocates to the nucleus in a GH-dependent manner, where it regulates the expression of other cell growth-related genes, and sensitises the cell to the proliferative action of GH.

Clinical Use of Growth Hormone to Enhance Cell Growth

The efficacy and safety of hGH therapy in promoting postnatal growth has resulted in its widespread use in many conditions of growth retardation. In addition to frank growth hormone (GH) deficiency, these applications include children born small for gestational age, children with chronic kidney disease, SHOX deficiency, Noonan syndrome, Prader-Willi syndrome, Turner syndrome, juvenile idiopathic arthritics on glucocorticoids, cystic fibrosis and children with idiopathic short stature (reviewed in [1]). Conversely, the GH receptor antagonist Pegvisomant, has proved efficacious in the treatment of giantism and acromegaly, where excess somatotrope secretion leads to excessive cell growth. Moreover, given the resistance of GH receptor-impaired individuals to cancer [2], it is plausible that GH antagonism will feature in future cancer therapies [3].

Growth Hormone and Cell Growth

The marked enhancement of cell growth by pituitary GH is evident postnatally in disorders of pituitary function, while the key role of the GH receptor (GHR) in this enhancement is evident in individuals lacking functional GHR (Laron dwarfism). The proportionality of normal postnatal growth has been attributed to the widespread actions of endocrine IGF-1 generated largely by GH action on the liver. Yet the finding of GHR in virtually all tissues and the finding that deletion of hepatic IGF1 has minimal effect on postnatal growth focused attention on local generation of IGF-1 as the growth mediator. The role of both locally produced IGF-1 and endocrine IGF-1 as key mediators of GH-dependent growth has received strong support from the ability of IGF-1 administration to promote respectable postnatal growth in individuals lacking functional GHR, although the extent of growth promotion is not to the extent seen with hGH treatment [1]. It was thought that this difference was a consequence of the rapid clearance of IGF-1 resulting from deficiency of GH dependent IGFBP3, but administration of a complex of IGF-1 and IGFBP3 proved to be inferior to IGF-1 alone. This review focuses on the ability of GH to promote cell growth both through its key role in generating IGF-1 and through its other direct genomic actions on most cells of the body.

Growth Hormone Receptor

The GHR was the first member of the class 1 cytokine receptors to be cloned, and this dismissed the classical view that the actions of GH were mediated through GH fragments [1, 4]. The GHR has proved to be paradigmatic for understanding signalling by members of what is now a receptor family of more than 25 members, including the receptors for erythropoietin, prolactin, leptin, thrombopoietin and most of the colony stimulating factors. These receptors function as homodimers or heteromers with auxillary subunits such as gp130, and use JAK tyrosine kinases to activate STAT transcription factors, resulting in nuclear translocation of tyrosine phosphorylated STAT dimers which regulate the transcription of a large number of genes. In addition, PI 3-kinase, ERK, JNK, NFkB and other signalling pathways are activated by these cytokine receptors [1].

In the case of the GHR, the homodimeric receptor activates JAK2 and a Src kinase which signal to several pathways, notably to STATs 1, 3 and 5, as well as the PI 3-kinase, ERK, NFkB, JNK and p38 kinase pathways (fig. 1).

Based on the use of targeted mutagenesis, cysteine crosslinking, FRET and molecular modelling, we have good evidence that the role of the GH ligand is to draw the two receptors together just above the transmembrane domain (TMD). Because the two receptor binding sites on GH are asymmetrically placed, binding of the two receptors results in relative rotation and vertical translocation of the two receptors. This results in a separation of the Box 1 JAK binding sequence below the TMD by a rotational movement of the TMDs, which facilitates activation of the two associated JAK2 kinases, resulting in STAT phosphorylation [1]. The growth signal is terminated by a combination of specific tyrosine phosphatase action (notably by PTP-H1

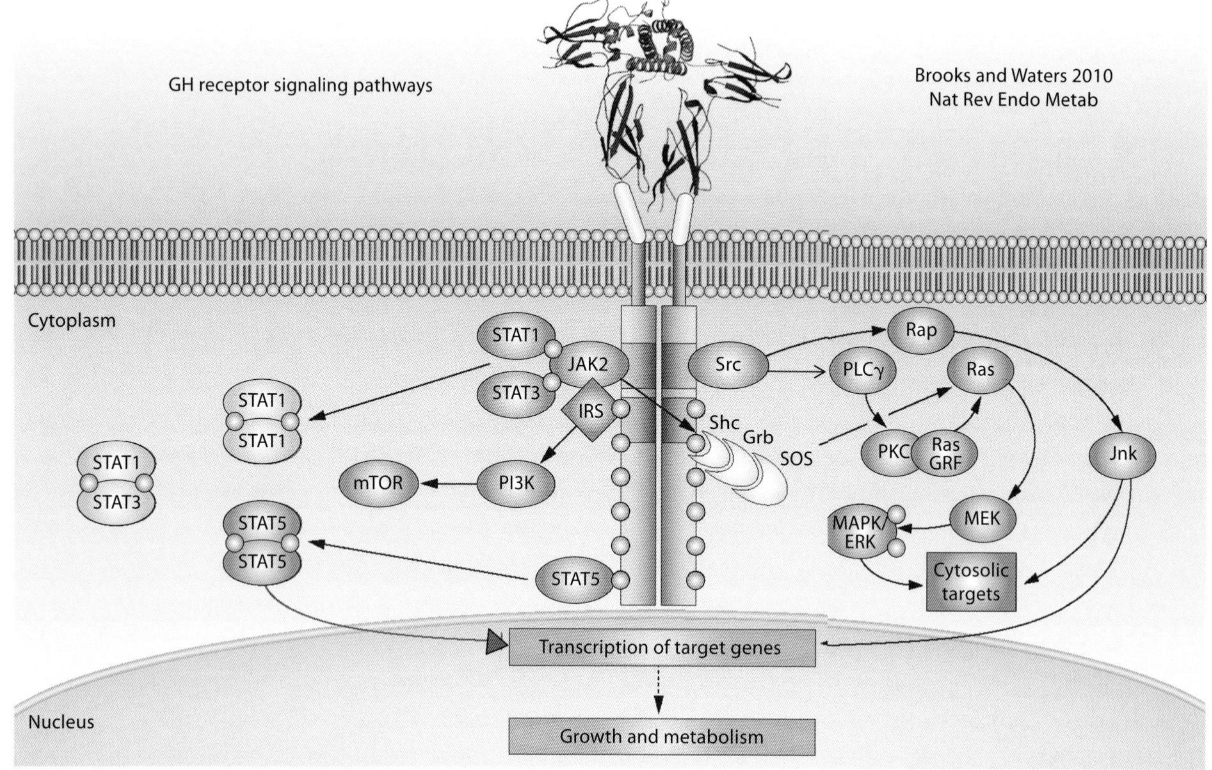

Fig. 1. Signalling pathways utilized by growth hormone. The extent of the individual pathways varies between cell types, depending on the relative expression of the component parts. Canonical protein-tyrosine kinase JAK2 signalling via STAT5 involves phosphorylation of key tyrosine residues in the receptor cytoplasmic domain (below residue 350), which bind the Src homology 2 (SH2) domain of STAT5A and STAT5B, recruiting these STATs to the activated JAK2 and thus facilitating their tyrosine phosphorylation and subsequent dimerization through their SH2 motifs. Dimerized STAT5 translocates to the nucleus to regulate gene transcription in complex ways. Conversely, STAT1 and STAT3 undergo direct tyrosine phosphorylation by JAK2 without the requirement for receptor binding. Erk can be activated either by Src and/or phospholipase Cγ and ras, or by JAK2 via the adaptors Shc, Grb and Sos. JNK is activated by Src via rap. The PI3K and the serine-threonine-protein kinase mTOR pathway is activated by JAK2 via IRS phosphorylation. Signals are initially damped or terminated by the action of specific tyrosine phosphatases and receptor internalization, then by the actions of induced SOCS proteins, particularly SOCS2. Erk = Extracellular signal-regulated kinase (also known as mitogen-activated protein kinase); Grb = growth factor receptor-bound protein; IRS = insulin receptor substrate; JAK2 = Janus kinase 2; JNK = c-Jun N-terminal kinase; MEK = dual specificity mitogen-activated protein kinase kinase 2; mTOR = mammalian target of rapamycin; PI3K = phosphoinositide 3-kinase; PLCγ = phospholipase Cγ; rap= ras-related protein; rasgrf = ras-specific guanine nucleotide-releasing factor; SHC = SH2-domain containing transforming protein; SOCS = suppressor of cytokine signaling; SOS = son of sevenless; Src = proto-oncogene tyrosine-protein kinase Src; STAT = signal transducer and activator of transcription. Reproduced from Brooks and Waters [1].

and PTP1B [5]) and the induction of SOCS (suppressors of cytokine signalling) proteins, notably SOCS2 [6].

In relation to postnatal growth, we know that STAT5 signalling is key to postnatal cell proliferation, since mice with knockin GHR mutations which prevent STAT5 phosphorylation by JAK2 are growth retarded to almost the same extent as mice completely lacking the GHR [7]. There is a difference of around 13% in GH-dependent growth between the STAT5-impaired *ghr391* mice and the full *ghr* knockout mice, and this difference is also seen when mice with deletion of ability to activate JAK2 (Box1 deleted mice) are examined [8]. Thus, we can conclude that, in mice, JAK-STAT5 signalling accounts for around 87% of GH-dependent growth, while direct JAK2 action accounts for the remainder. This remainder is similar to the IGF-1-independent growth by GH which is seen when comparing *igf-1* knockout mice with combined *igf-1* and *ghr* knockout mice, where the double knockout mice have a final weight of only 17% of wild type, whereas the *igf-1* knockout alone is 30% of wild type [9]. What could account for this IGF-1-independent postnatal growth of around 13%? The fact that preventing activation of JAK2 (Box1 deleted mice) is as effective as full GHR knockout tells us that Src-ERK signalling does not separately contribute to postnatal growth, although it does have an important role in immunoprotection during liver regeneration. Hence the residual growth is GHR-JAK2 dependent, but GHR-STAT5 independent, which appears to rule out induction of EGF, prolactin and estrogen alpha receptors, all GHR-STAT5 dependent [7]. Likely candidates are the Akt/PI 3-kinase pathway and the NFkB pathway resulting in increased proliferation and decreased apoptosis mediated via Bcl-2 [10], as well as JAK-STAT3 induction of c-myc expression [11].

It should be noted that while STAT5-induced IGF-1 expression is mandatory for most GH-dependent postnatal cell growth, STAT5 is a transcription factor with many other targets, and these include key mitogenic and survival factors such as cyclin D1 and Bcl-xL [12], and these presumably synergise with IGF-1 in promoting cell growth, together with STAT5-induced growth factor receptors such as EGFR, PrlR, ERα and AR. Moreover, STAT5 directly associates with several key regulatory proteins such as the GR and TGFR which may enhance IGF-1-mediated proliferation [3].

Unique Importance of STAT5 for Postnatal Growth

While the above mouse models correlate loss of ability of the GHR to activate STAT5 with loss of most GH-dependent postnatal growth, they do not unequivocally demonstrate a role for STAT5, since other signalling elements could be lost with the GHR391 truncation. The demonstration that STAT5b is critical came from clinical studies identifying a number of severely growth retarded individuals lacking functional STAT5b. The first of these [13] coincided with the reporting of functional STAT5 elements in the *igf-1* gene promoter by Rotwein's group [14]. We now know that there are at least

7 groups of these transcriptional enhancer elements, which together confer strong STAT5 dependence on *igf-1* gene expression, thus underpinning GH-dependent cell growth [14].

We can visualize STAT5 activation (phosphorylation) in the growth plate by immunohistochemistry, using a phospho-STAT5 antibody. A comparison of wild type and GH deficient *dw/dw* rats shows striking differences in phospho-JAK2 and phospho-STAT5 in the reserve and prehypertrophic zones of the tibial growth plate, with phospho-STAT5 immunoreactivity correlating with GHR immunoreactivity [15]. GH injection increased phospho-STAT5 in these zones rapidly. This is consistent with GH stimulation of resting chondrocytes, so increasing the number of IGF1 producing proliferative and hypertrophic chondrocytes. No stimulation of phospho-STAT5 was seen with either prolactin or IGF-1 administration. Importantly, fasting for 48 h markedly impaired the phospho-STAT5 response to administered GH, in accord with the loss of growth in starvation [15].

Given the role of GHR activation in STAT5 activation, and the role of this transcription factor in generation of IGF-1, we would expect to find GHR and IGF-1 co-localizing in most tissues. This is often the case, such as in the formation of hard tissues like the tooth and bone [16], and in myocytes, hepatocytes, and adipocytes, but in the chondrocytes of the growth plate the IGF-1 appears to be acting more in a paracrine or endocrine mode than an autocrine mode, since GHR immunoreactivity is only weak or absent in proliferating and prehypertrophic chondrocytes where IGF-1 immunoreactivity and transcript occur [17], and these zones do not respond to GH with STAT5 phosphorylation [15]. Presumably, a significant component of the IGF-1 immunoreactivity is derived from endocrine IGF-1, since around 30% of postnatal growth is now reported to be driven by endocrine IGF-1 in the mouse [18].

Growth Hormone and Stem Cell Recruitment

The localization of GH responding cells to the reserve zone and Groove of Ranvier in the growth plate suggests a role for GH in activation of stem cells. Based on the ability of GH but not IGF-1 to increase chondrocyte colony size in vitro [19], this would appear to be the case, and a requirement for GH but not IGF-1 to replenish chondrocyte stem cell number may well account for the more marked growth promoting activity of hGH compared to IGF-1. Supporting this view, GHR null mice have impaired chondrocyte proliferation, in contrast to IGF-1 null mice which have impaired chondrocyte hypertrophy [20]. Hence the key role of GH in long bone growth appears to be its ability to recruit chondrocyte stem cells.

That GH is able to promote stem cell activation and proliferation is clearly demonstrated in the case of neural stem cells. We have demonstrated GHR expressing neural stem cells (NSC) in mice which express the full-length GHR, rather than the alternatively spliced or cleaved GH-binding protein [21, 22]. These NSCs are multipotent and capable of forming the three neural cell types in vitro (neurons, oligodendrocytes and astrocytes), and respond to GH, but not to IGF-1 with neurogenesis in vivo (i.e. formation of new neurons in the olfactory bulb). Further, total NSC number is decreased in GHR null mice, and conversely, NSC number is increased by GH infusion into the cerebral ventricles of wild type mice. Importantly, while NSC number declines with age (as does GH secretion and IGF-1), this decline can be reversed by either voluntary exercise (running) or by rat GH infusion into the cerebral ventricles. We found that the exercise-induced increase in NSCs does not occur in GHR deleted mice or GHR391 mice, unable to activate STAT5. These findings provide evidence for a new physiological role for GH, ie promoting neural stem cell proliferation [23]. This role is concordant with meta-analyses showing impaired cognitive function in clinical GH deficiency (reviewed in [24]), together with altered motor function and brain structure. It is also concordant with findings of altered cortical structure in GHR null mice as well as decreased brain volume in GHR null humans [24].

Growth Hormone Receptor in the Cell Nucleus and Its Role in Cell Proliferation

We have reported a high correlation between hepatocyte replication and the presence of GHR in the nucleus during liver regeneration in the rat [25]. Nuclear GHR is also present in highly proliferative cells in the GIT, placenta and growth plate, and in preimplantation embryos, suggesting a role in cell proliferation. The nuclear GHR is full-length based on confocal Z-scan microscopy with N- and C-terminal epitope tags [25]. In BaF/B03 cells expressing GHR, the addition of GH to serum starved cells caused rapid (within 30 min) translocation of the GHR to the nucleus. We showed that Importin-beta is able to bind to the extracellular domain of the GHR, and therefore could facilitate the nuclear translocation, and addition of antibodies to Importin-beta prevented nuclear translocation in an in vitro model. While the exact mechanism is not resolved, extensive study of the EGFR, which also nuclear translocates in response to ligand binding, has shown that it transits through the nuclear pore and exits from the inner nuclear membrane into the nucleoplasm through Sec67 retrograde transport [26]. Nuclear translocation of a number of other tyrosine kinase receptors has been documented, including ErbB2, FGFR1 and c-Met tyrosine kinase receptors.

What is a membrane receptor and its ligand doing in the nucleus? In some cases the receptor is thought to act via its tyrosine kinase domains in a more conventional manner, whereas in others the receptor has been shown to interact directly with the DNA as a transcriptional activator. The former is exemplified by nuclear-targeted FGFR1, which is able to initiate c-jun transcription, but only when its tyrosine kinase activity is present, and by EGFR which phosphorylates tyrosine 211 on chromatin-bound PCNA (proliferating cell nuclear antigen) and stabilizes it, stimulating DNA replication. Direct interaction of the nuclear receptor with the genome is also exemplified by EGFR, which binds to and activates an AT-rich sequence in the *cyclin D1, iNOS, aurora-A* and *B-myb* promoters [26], and ErbB2, which similarly transactivates the COX-2 promoter. Importantly, both in the case of the EGFR and the FGFR, there is a correlation between the presence of the receptor in the nucleus and cell division (e.g. in liver regeneration, and in both cases there is ligand-induced transport of receptor to the nucleus).

We sought to define roles of the nuclear GHR by forced nuclear localization with a nuclear localization sequence (NLS). Cells stably expressing similar amounts of NLS-GHR and Wt-GHR showed a major difference in sensitivity to GH, with the dose curve for cell proliferation shifting 20-fold leftwards with the NLS-GHR. This was associated with upregulation of a set of proliferation related genes including *survivin, mybbp1, cathepsin C, phgdh* and *dysadherin*, as determined by unbiased microarray analysis. Our mechanistic studies then showed that nuclear localized GHR results in constitutively activated STAT5 associating with GHR in the nucleus, and this was partly a result of increased sensitivity to autocrine GH [25].

Proteins Interacting with Extracellular Domain of Nuclear GHR

In an effort to further investigate the mechanism of action of nuclear GHR, we sought interactors with the extracellular domain of the receptor using yeast 2-hybrid methodology. We chose the ECD/GHBP since we had earlier found this to be present in the nucleus [27]. To our surprise, the ECD bait was a transcriptional activator, and this property was mapped to the lower cytokine receptor domain. Transcriptional activation activity was abolished when Ser226 of the conserved cytokine receptor WSxWS sequence was mutated to alanine. We then used affinity chromatography of nuclear extracts with immobilized extracellular domain of the receptor (GHBP), to identify interactors, since we could not do this in the yeast 2 hybrid assay because of transcriptional activation by the GHBP bait. As a control we used immobilized S226A GHR ECD, which was not transcriptionally active in CHO cells and yeast. Using SDS-PAGE and mass spectroscopy, we were able to identify a number of interacting proteins, including a nucleoporin (Nup145) and two RNA binding proteins – one an important coactivator and splicing protein implicated in cancer (CoAA/SIP).

CoAA (also known as SIP because it interacts with the *syt* oncogene for chondrosarcoma) was particularly interesting because it is a powerful nuclear hormone coactivator and splicing protein with RRMs (RNA recognition motifs), and is chromosomally amplified in lymphoma and squamous cell carcinoma. CoAA is a ubiquitous protein which interacts with TRBP (thyroid hormone receptor binding protein/ nuclear hormone receptor coregulator), enhancing polymerase II-dependent transcription. Having identified the interaction of GHR with CoAA by affinity chromatography, we then established by CoIP that CoAA bound to the receptor only after GH treatment of cells, and that this binding required Ser226. Importantly, transfection of

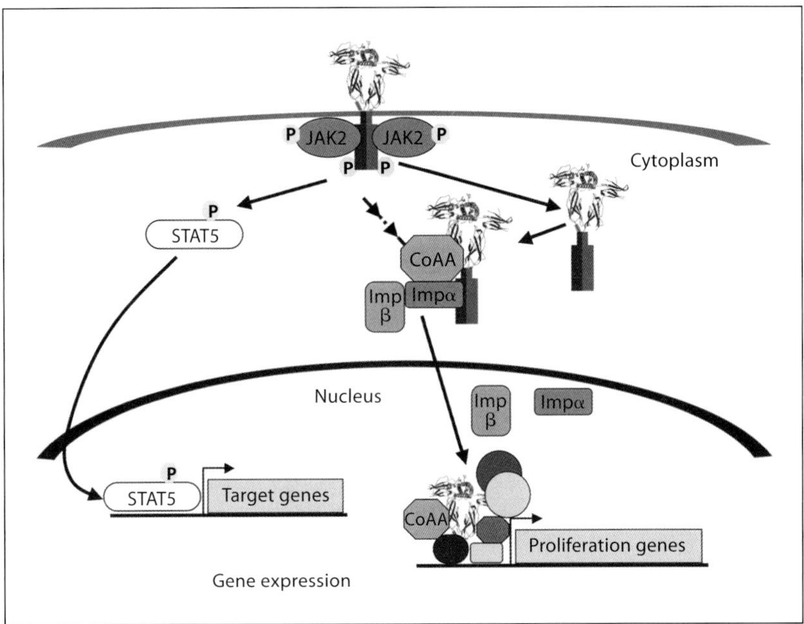

Fig. 2. Model of proposed mechanism for nuclear GHR action. Plasma membrane localized GHR induces STAT5 activation, and other pathways, by the classical JAK-STAT pathway. STAT5 translocates to the nucleus and binds to its DNA recognition sequence to initiate target gene transcription. In certain cellular conditions, such as during the proliferative phase in liver regeneration and at defined cell cycle-regulated times in progenitor proliferative cells, the GHR escapes its degradative pathway after GH addition and is translocated to the cytoplasm (possibly via the sec61 translocon) and then to the nucleus by the importin-α/β mediated classical import pathway. This process may involve its interaction with the NLS-containing protein CoAA, which mediates the interaction between GHR and nuclear import machinery. Once in the nucleus, the GHR can act as a transcriptional activator in conjunction with CoAA to initiate transcription of a subset of target genes to regulate cell cycle progression. Imp = Importin; JAK = Janus family of tyrosine kinases. Reproduced from Conway-Campbell et al. [27], with permission. Copyright 2008, The Endocrine Society.

CoAA into cells rendered them more sensitive to GH stimulation of proliferation (doubled the maximum response to GH without increasing GH receptor expression), although it did not affect mitogenic stimulation by IL-3 in these lymphoid BaF/B03 cells. Presumably the association of CoAA with the receptor extracellular domain is part of the enhanced sensitivity to GH seen when the receptor is nuclear localized. The proposed role of nuclear GHR is shown in figure 2.

Conclusions

GH action accounts for approximately half of postnatal growth, and this is largely mediated through activation of STAT5. This transcription factor is necessary for the mandatory role of IGF-1 induction in growth, but by virtue of its target gene promiscuity regulates many other proliferation-related genes. As well as inducing IGF-1, GH is also able to directly activate stem cells in the brain and in the growth plate. Cell sensitivity to GH is enhanced by nuclear localization of the receptor in a GH-dependent manner, and the nuclear GHR recruits a powerful co-activator, CoAA to assist in sensitizing the cell to the growth actions of GH.

References

1 Brooks AJ, Waters MJ: The growth hormone receptor: mechanism of activation and clinical implications. Nat Rev Endocrinol 2010;6:515–525.

2 Guevara-Aguirre J, Balasubramanian P, Guevara-Aguirre M, Wei M, Madia F, Cheng CW, Hwang D, Martin-Montalvo A, Saavedra J, Ingles S, de Cabo R, Cohen P, Longo VD: Growth hormone receptor deficiency is associated with a major reduction in pro-aging signaling, cancer, and diabetes in humans. Sci Transl Med 2011;3:70ra13.

3 Chhabra Y, Waters MJ, Brooks AJ: Role of the GH/IGF1 axis in cancer. Expert Rev Endocrinol Metab 2010;6:71–84.

4 Rowlinson SW, Waters MJ, Lewis UJ, Barnard R: Human GH fragments 1–43 and 44–191: in vitro somatogenic activity and receptor binding characteristics in human and nonprimate systems. Endocrinology 1996;137:90–95.

5 Pilecka I, Patrignani C, Pescini R, Curchod ML, Perrin D, Xue Y, Yasenchak J, Clark A, Magnone MC, Zaratin P, Valenzuela D, Rommel C, Hooft van Huijsduijnen R: Protein-tyrosine phosphatase H1 controls growth hormone receptor signaling and systemic growth. J Biol Chem 2007;282:35405–35415.

6 Greenhalgh CJ, Rico-Bautista E, Lorentzon M, Thaus AL, Morgan PO, Willson TA, Zervoudakis P, Metcalf D, Street I, Nicola NA, Nash AD, Fabri LJ, Norstedt G, Ohlsson C, Flores-Morales A, Alexander WS, Hilton DJ: SOCS2 negatively regulates growth hormone action in vitro and in vivo. J Clin Invest 2005;115:397–406.

7 Rowland JE, Lichanska AM, Kerr LM, White M, d'Aniello EM, Maher SL, Brown R, Teasdale RD, Noakes PG, Waters MJ: In vivo analysis of growth hormone receptor signaling domains and their associated transcripts. Mol Cell Biol 2005;25:66–77.

8 Barclay JL, Kerr LM, Arthur L, Rowland JE, Nelson CN, Ishikawa M, d'Aniello EM, White M, Noakes PG, Waters MJ: In vivo targeting of the growth hormone receptor (GHR) Box1 sequence demonstrates that the GHR does not signal exclusively through JAK2. Mol Endocrinol 2010;24:204–17.

9 Lupu F, Terwilliger JD, Lee K, Segre GV, Efstratiadis A: Roles of growth hormone and insulin-like growth factor 1 in mouse postnatal growth. Dev Biol 2001; 229:141–62.

10 Jeay S, Sonenshein GE, Postel-Vinay MC, Kelly PA, Baixeras E: Growth hormone can act as a cytokine controlling survival and proliferation of immune cells: new insights into signaling pathways. Mol Cell Endocrinol 2002;188:1–7.

11 Kiuchi N, Nakajima K, Ichiba M, Fukada T, Narimatsu M, Mizuno K, Hibi M, Hirano T: STAT3 is required for the gp130-mediated full activation of the c-myc gene. J Exp Med 1999;189:63–73.

12 de Groot RP, Raaijmakers JA, Lammers JW, Koenderman L: STAT5-Dependent CyclinD1 and Bcl-xL expression in Bcr-Abl-transformed cells. Mol Cell Biol Res Commun 2000;3:299–305.

13 Hwa V, Nadeau K, Wit JM, Rosenfeld RG: STAT5b deficiency: lessons from STAT5b gene mutations. Best Pract Res Clin Endocrinol Metab 2011;25: 61–75.

14 Chia DJ, Varco-Merth B, Rotwein P: Dispersed Chromosomal Stat5b-binding elements mediate growth hormone-activated insulin-like growth factor-I gene transcription. J Biol Chem 2010;285: 17636–17647.

15 Gevers EF, Hannah MJ, Waters MJ, Robinson IC: Regulation of rapid STAT5 phosphorylation in the resting cells of the growth plate and in liver by GH and feeding. Endocrinology 2009;150:3627–3636.

16 Joseph BK, Savage NW, Young WG, Guota GS, Breier B, Waters MJ: Expression and regulation of insulin-like growth factor-1 in the rat incisor. Growth Factors 1993;8:267–275.

17 Parker EA, Hegde A, Buckley M, Barnes KM, Baron J, Nilsson O: Spatial and temporal regulation of GH-IGF-related gene expression in growth plate cartilage. J Endocrinol 2007;194:31–40.

18 Stratikopoulos E, Szabolcs M, Dragatsis I, Klinakis A, Efstratiadis A: The hormonal action of IGF1 in postnatal mouse growth. Proc Natl Acad Sci USA 2008;105:19378–19383.

19 Isaksson OG, Lindahl A, Nilsson A, Isgaard J: Mechanism of the stimulatory effect of growth hormone on longitudinal bone growth. Endocr Rev 1987;8:426–438.

20 Wang J, Zhou J, Cheng CM, Kopchick JJ, Bondy CA: Evidence supporting dual, IGF-I-independent and IGF-I-dependent, roles for GH in promoting longitudinal bone growth. J Endocrinol 2004;180: 247–255.

21 Blackmore DG, Reynolds BA, Golmohammadi MG, Large B, Aguilar RM, Haro L, Waters MJ, Rietze RL: Growth hormone responsive neural precursor cells reside within the adult mammalian brain. Sci Rep 2012;2:250.

22 Harrison SM, Barnard R, Ho KY, Rajkovic I, Waters MJ: Control of GH binding protein release from human hepatoma cells expressing full length GH receptor. Endocrinology 1995;136:651–659.

23 Blackmore DG, Golmohammadi MG, Large B, Waters MJ, Rietze RL: Exercise increases neural stem cell number in a growth hormone-dependent manner, augmenting the regenerative response in aged mice. Stem Cells 2009;27:2044–2052.

24 Waters MJ, Blackmore DG: Growth hormone (GH), brain development and neural stem cells. Pediatr Endocrinol Rev 2011;9:549–553.

25 Conway-Campbell BL, Wooh JW, Brooks AJ, Gordon D, Brown RJ, Lichanska AM, Chin HS, Barton CL, Boyle GM, Parsons PG, Jans DA, Waters MJ: Nuclear targeting of the growth hormone receptor results in dysregulation of cell proliferation and tumorigenesis. Proc Natl Acad Sci USA 2007;104: 13331–13336.

26 Wang YN, Yamaguchi H, Huo L, Du Y, Lee HJ, Lee HH, Wang H, Hsu JM, Hung MC: The translocon Sec61beta localized in the inner nuclear membrane transports membrane-embedded EGF receptor to the nucleus. J Biol Chem 2010;285:38720–38729.

27 Conway-Campbell BL, Brooks AJ, Robinson PJ, Perani M, Waters MJ: The extracellular domain of the growth hormone receptor interacts with coactivator activator to promote cell proliferation. Mol Endocrinol 2008;22:2190–2202.

Michael J. Waters
Institute for Molecular Bioscience, The University of Queensland
St. Lucia, QLD 4072 (Australia)
Tel. +61 7 3346 2037, E-Mail m.waters@uq.edu.au

Mullis P-E (ed): Developmental Biology of GH Secretion, Growth and Treatment.
Endocr Dev. Basel, Karger, 2012, vol 23, pp 96–108 (DOI: 10.1159/000341763)

From Endoplasmic Reticulum to Secretory Granules: Role of Zinc in the Secretory Pathway of Growth Hormone

Vibor Petkovic · Maria Consolata Miletta · Primus-E. Mullis

University Children's Hospital, Pediatric Endocrinology, Diabetology and Metabolism, Inselspital, Bern, Switzerland

Abstract

Endocrine and neuroendocrine cells differ from cells which rapidly release all their secreted proteins in that they store some secretory proteins in concentrated forms in secretory granules to be rapidly released when cells are stimulated. Protein aggregation is considered as the first step in the secretory granule biosynthesis and, at least in the case of prolactin and growth hormone, greatly depends on zinc ions that facilitate this process. Hence, regulation of cellular zinc transport especially that within the regulated secretory pathway is of importance to understand. Various zinc transporters of Slc30a/ZnT and Slc39a/Zip families have been reported to fulfil this role and to participate in fine tuning of zinc transport in and out of the endoplasmic reticulum, Golgi complex and secretory granules, the main cellular compartments of the regulated secretory pathway. In this review, we will focus on the role of zinc in the formation of hormone-containing secretory granules with special emphasis on conditions required for growth hormone dimerization/aggregation. In addition, we highlight the role of zinc transporters that govern the process of zinc homeostasis in the regulated hormone secretion.

All eukaryotic cells have the route for intracellular transport of newly synthesized proteins, generally referred to as the secretory pathway, which they use for the transport of secretory and membrane proteins to the plasma membrane or to lysosomes. In many of them, secretory proteins are transported as soon as they are synthesized through the constitutive secretory pathway and their secretion by exocytosis that does not require stimulation. Endocrine and neuroendocrine cells have an additional specialized pathway (the regulated secretory pathway) in which proteins are 'packed' in concentrated forms in membrane-enclosed core vesicles with dense cores also called secretory granules. Upon specific stimulation, these vesicles fuse with the plasma membrane causing the dense cores to be released and dissolved leading to a burst of hormone in the bloodstream. Basically, proteins destined for secretion are synthesized as larger precursors on the rough endoplasmic reticulum (ER) and immediately transported through a pore recognizing the signal peptide in the ER membrane into the luminal spaces where they are properly folded by ER-resident chaperons [1] and subsequently transported to the Golgi complex. Transport of most secretory proteins through the Golgi complex follows the model called cisternal maturation [2–4], which starts with arrival of COPII vesicles carrying soluble proteins from the lumen of the ER at the intermediate compartment where they fuse to form a new Golgi cisterna. As the newly formed cisterna moves up through the Golgi stacks starting from the *cis-* and progressing further to the medial- and to *trans-*Golgi side, newer cisternae from behind it and older ones ahead mature and are eventually disassembled [5]. During this 'trip' through the Golgi stacks, the Golgi cisterna matures and many proteins targeted for secretion localized in the lumen undergo series of enzymatic modifications that include phosphorylation, sulfation and glycosylation in the case of glycoproteins. In addition, a number of proteins are processed from their pro-forms to their mature forms. Furthermore, the lumen of *trans-*Golgi layer is a particular place within the secretory pathway where aggregation of some hormone proteins occurs (discussed more in detail below) which has been considered as one of the crucial steps in the formation of secretory granules. The last two to three *trans-*Golgi layers, known as the *trans-*Golgi network (TGN), serve as one of the 'sorting stations' within the secretory pathway where different secretory proteins are sorted into distinct tubular carriers that are targeted to different final destinations like plasma membrane, lysosomes, early, late and recycling endosomes and secretory granules [5]. Once hormone aggregate is formed, specific membrane proteins necessary for transport and release of secretory granules recognize aggregate 'content' and become localized in the membrane around aggregate. Excess of membranes, inappropriate membrane and other soluble proteins are removed through a number of maturation steps that progressively convert secretory granules into their mature form ready to be secreted via triggering by a secretagogue [6]. These include acidification of immature secretory granules to activate enzymes for the processing of the regulated secretory pathway proteins, removal of constitutive secretory proteins and lysosomal enzymes packed into immature secretory granules, loss of clathrin coat and other coat proteins and condensation of their protein constituents increasing secretory granule dense cores. Once formed, mature secretory granules retain their identity throughout the exo-endocytic cycle.

Human growth hormone (GH) follows the same maturation process after being produced in the major endocrine cell type in the anterior pituitary gland, the somatotropes. GH is a globular, monomeric protein with a molecular mass of 22 kD, which is primarily produced as a precursor of 217 amino acids (aa). In order to enter into the ER, a signal peptide (26 aa) is enzymatically cleaved off yielding a single polypeptide chain of 191 aa residues with two disulfide bridges and no carbohydrate moieties. GH is than stored in concentrated forms in secretory granules enabling a regulated release upon GHRH stimulation [7, 8]. Furthermore, the histochemical analysis of the anterior pituitary gland indicated that zinc ions (Zn^{2+}) are present in high concentrations in the Golgi complex and GH-containing secretory granules [9]. This review outlines the role of Zn^{2+} in secretory granule biosynthesis, with the emphasis on its role in dimerization/aggregation of GH required for normal storage within secretory granules. Moreover, it focuses on the regulation of Zn^{2+} transport by various zinc transporters and especially on their role in mobilization of Zn^{2+} in the regulated secretory pathway.

Protein Aggregation in the Formation of Secretory Granules

Spherical in shape with diameters of several hundred nanometers, secretory granules serve as a storage pool for proteins selected for secretion. The complex process of secretory granule biogenesis begins with aggregation of proteins destined for secretion to form dense cores of granules composed of large insoluble aggregates that form by self-association. Aggregation of secretory proteins as such, is thermodynamically favourable process since it does not require energy (a passive process). In addition, it also provides an efficient method of protein sorting, because the size of aggregates alone is enough to exclude them from other pathways mediated by tubular export carriers. After being released from cells by exocytosis, these aggregates are rapidly solubilized into monomeric forms with their native conformations. The process of aggregation, as an early step in the formation of secretory granules, has been quite extensively investigated in the case of human prolactin (PRL) and based on the structural similarities, there are good predictions that human GH might behave in the similar manner (discussed more in detail below).

In order to investigate process of hormone aggregation in cells, Lee et al. [10] developed an assay based on previously reported findings that membranes of PRL-containing secretory granules isolated from rat anterior pituitary glands can be dissolved in Lubrol (nonionic detergent) leaving the dense cores (protein aggregates) of PRL granules as Lubrol-insoluble aggregates that can be isolated by sedimentation [11]. This allows the solubility of newly synthesized hormones to be followed with time using pulse-chase techniques (usually done with ^{35}S-labeled amino acids). The experiments were performed in rat pituitary tumor cells (GH_4C_1) that endogenously produce and secrete PRL and GH. The data demonstrated that PRL is soluble in Lubrol immediately after synthesis while 30 min later roughly 40–50% of newly synthesized hormone acquires Lubrol-insoluble forms [10]. Furthermore, when transport from the ER is prevented by brefeldin A or incubation of the cells at 15°C, the conversion of 'labeled' PRL to an insoluble form was inhibited indicating that aggregation which occurs within 30 min after synthesis requires transport from the ER within that period of time.

There is protein specificity to the process of protein aggregation, which in other words means that not all secretory proteins aggregate in pituitary cells. For instance, human serum albumin normally secreted from the liver without aggregation, displays quite low degree of aggregation (less than 10%) having no tendency to increase it with time when transiently expressed in AtT-20 (mouse pituitary cells that store ACTH in secretory granules) [10].

A specificity of secretory protein aggregates and degree of their retention in cells is interesting to observe in the case of GH-R183H mutant, reported to cause autosomal dominant growth hormone deficiency [12]. An individual carrying this mutation has releasable stores of GH and results of different stimulation tests indicate GH secretion to be severely impaired. After transient transfection of AtT-20 cells with *wt*-GH or GH-R183H, both GH variants are synthesized in the same amount, both form equivalent amounts of insoluble aggregates after synthesis and consequently both are stored in secretory granules. Interestingly, despite the fact that GH-R183H protein folds correctly and has full bioactivity, the mutant differed from *wt*-GH in its retention in the cells after packaging into secretory granules (50% more GH-R183H retained in AtT-20 cells compared to *wt*-GH). Furthermore, the analysis of stimulated secretion in AtT-20 cells singly expressing each of the variants revealed decreased secretion of GH-R183H when compared to *wt*-GH and dominant effect of the mutant partially suppressing secretion of the *wt*-GH in co-transfection experiments [13]. This makes the GH-R183H a good example of how a single amino acid change in the dense cores of granules can apparently affect granule behavior.

Protein aggregation takes place in the lumen of *trans*-Golgi layer where specific environmental factors like pH and higher concentrations of divalent ions like zinc are thought to trigger this process. The study of pH regulation in live AtT-20 cells demonstrated that steady-state pH decreased from 7.2 in the ER to 6 in the Golgi and 5.5 in mature secretory granules [14] making these data in line with generally accepted model that aggregation occurs in *trans*-Golgi compartment at acidic pH. Aggregation of rat PRL in GH_4C_1 cells and human GH expressed in neuroendocrine AtT-20 cells was slowed but not completely prevented by agents neutralizing intracellular acidic compartments [10, 15]. On the other hand, aggregation of human PRL in AtT-20 cells was completely prevented leading to the conclusion that aggregation of PRL is more dependent on acidic pH than that of GH [15]. Apart from specific pH requirements, protein aggregation apparently requires high amounts of divalent cations like zinc ions (Zn^{2+}) since aggregation of human PRL expressed in AtT-20 cells was slowed by incubation with diethylcarbamate, a membrane-permeable chelator of Zn^{2+} [15].

One approach for investigating the mechanisms and principles of protein aggregate formation at the molecular level which has been widely accepted is to study these events in solution. In order to mimic what happens in cells, the following specific conditions must be fulfilled. These include formation of aggregates which must occur within 30 min after synthesis in conditions that exist in the *trans*-Golgi lumen (acidic pH and high concentration of Zn^{2+}) and rapid dissociation of aggregates to native hormone in conditions found outside the cells. In addition, macromolecular crowding is also an important factor which should be considered, but it is often underestimated.

In the cytoplasm of cells, macromolecules take up to 30% of volume while the concentration of proteins has been estimated to be 200–300 mg/ml [16]. This reduces the volume available for reactions like protein aggregation leading to an increase in the thermodynamic activity of a protein, which in this case, would self-associate making an aggregate [17]. The lumen of the secretory pathway is likely to be as crowded as the cytoplasm itself. By adding large inert molecules in solution, such as polyethylene glycol (PEG) or dextran, it is possible to mimic crowded conditions found in cells (cytoplasm). The ability of human PRL to aggregate and the properties of the aggregate have been extensively studied in crowding vs. noncrowding conditions [15]. In crowding conditions PRL aggregates at pH 6 (like in the *trans*-Golgi lumen) while aggregation is not observed at pH 7.4 (similar to that of the ER) being consistent with its behavior in cells. Furthermore, molecular crowding changes the pH dependence of PRL interaction with Zn^{2+} in a way that it occurs more readily in an acidic milieu.

Zn^{2+} binding to human PRL and GH occurs through high-affinity binding sites (histidine residues) [18, 19] or through a low-affinity association with amino acids like glutamate, aspartate and glutamine [20]. However, H27A-PRL, the mutant of human PRL reported not to bind Zn^{2+} [19], forms aggregates in the presence of Zn^{2+} displaying a similar degree of aggregation as the *wt*-PRL [15].

Taken together, a high concentration of PRL and Zn^{2+} combined with acidic pH in the *trans*-Golgi lumen cause PRL to self-associate making oligomers. At acidic pH, histidine residues in PRL are protonated which prevents their binding to Zn^{2+} [20]. Hence, Zn^{2+} may facilitate aggregation of PRL by binding to glutamate and aspartate residues (low-affinity binding) and cross-bridging small aggregates into larger ones helping in this way, the formation of PRL dense cores in secretory granules. Finally, as mentioned earlier similar mechanism might be true for GH.

Dimerization versus Oligomerization of GH by Zinc: which Is the One that Mediates GH Storage in Secretory Granules?

Zinc is the second most abundant 'trace' element in the body and it is vital for normal cellular function playing a structural role in transcription factors and related proteins as well as structural and catalytic roles in numerous secretory, membrane-bound, and endosome/lysosome-resident enzymes [21–23]. Over two decades ago, a high concentration of Zn^{2+} was reported to be localized mostly in secretory granules and to lesser extent in Golgi complex of rat anterior pituitary cells [9], thus confirming its presence in the regulated secretory pathway. Nowadays, by using advanced microscopy techniques (confocal microscopy), we can visualize intracellular Zn^{2+} using a specific dye, ZP-1 (zinpyr-1), that has a nanomolar affinity for Zn^{2+}. AtT20 cells expressing GH fused to the Cherry fluorescent protein were additionally incubated in saline solution containing ZP-1 (fig. 1). Confocal imaging of ZP-1 fluorescence (Zinc) and fluorescence of GH-cherry fluorescent protein (GH-Cherry) revealed the expression pattern of GH

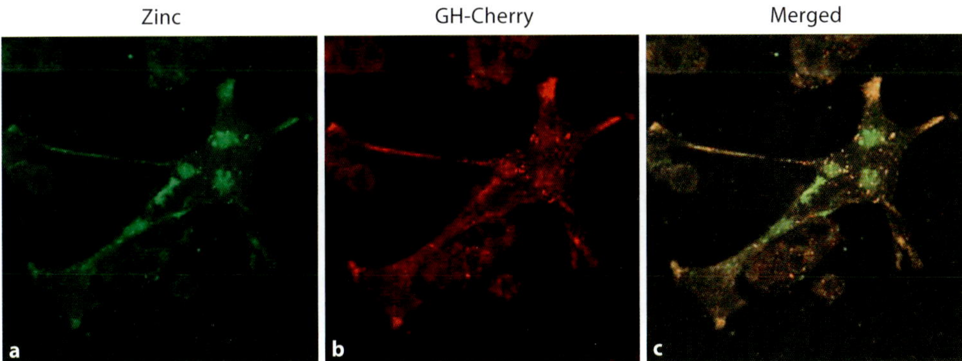

Fig. 1. Zinc localization with GH in AtT-20 cells using ZP-1 dye. AtT-20 cells (mouse pituitary cell line) were transfected with GH-Cherry (GH-RFP) using Lipofectamine 2000 (Invitrogen AG). 24 h after transfection, media was changed to fresh media without FCS and cells were incubated further for 2–4 h. Media was then replaced with a saline solution containing 2 mM ZP-1 for 20 min and washed for at least 30 min prior to confocal imaging (BioRad Radiance 2000). Imaging was done using an excitation wavelength of 488 nm, capturing the fluorescence between 515 and 535 nm for Zn^{2+} (**a**) and excitation wavelength of 543 nm capturing the fluorescence between 550 and 590 nm for GH-Cherry (**b**). Color colocalization (**c**) was achieved by merging the image input from the two respective channels.

and intracellular distribution of Zn^{2+} in a single cell. Color colocalization (merged) is the result of an overlap of both channel inputs from a single cell. Hence, the merge color (yellow) in punctate staining near the plasma membrane might be consistent with accumulation of GH peptide and high Zn^{2+} content in secretory granules.

At present, there is enough evidence to support the hypothesis that such unusually high concentrations of Zn^{2+} localized in secretory granules participates in the process of secretory granule biogenesis by means of facilitating protein aggregation. The first clues pointing towards this direction came in the late 1980s with the report describing that the mutant variant of human insulin (Asp-B10-insulin), which presumably does not bind Zn^{2+}, cannot form hexamers considered to be important for the formation of insulin crystals. Thus, these mutants are not properly processed throughout the secretory pathway presumably due to the inability to be concentrated in secretory granules [24]. As already mentioned in the previous section, Zn^{2+} binding to human prolactin (PRL), which is from the structural point of view closely related to GH [25], was reported to be crucial for its intracellular processing and normal secretion [19].

The first implications of Zn^{2+} involvement in GH storage in secretory granules came with the study reporting that Zn^{2+} is required for dimerization of GH, as demonstrated by size-exclusion chromatography and sedimentation equilibrium analysis [18]. In fact, two Zn^{2+} are believed to associate per GH dimer in a cooperative fashion through binding at high-affinity residues in GH (His18, His21 and Glu174) and mutation of these residues to alanine caused reductions of dimeric GH formation. Importantly, in this study it was also demonstrated that Zn^{2+} binding to GH would enhance the stability of the stored form (GH denaturation measured by guanidine hydrochloride at pH 8) and that the Zn^{2+}-GH complex was more stable to denaturation when compared to monomeric GH [18]. In order to investigate further whether disturbed Zn^{2+} binding of GH mutants (in which single, double or triple alanine substitutions at positions His18, His21 and Glu174 have been introduced) inhibits *wt*-GH secretion or production or if *wt*-GH and/or different GH mutants were transiently expressed in GH_4C_1 cells [26]. The measurement of constitutive GH secretion (i.e. without stimulation) and intracellular production (both measured by RIA) in culture media and cell lysates after 48 h of incubation revealed that both *wt*-GH secretion and production remained unaffected when *wt*-GH was coexpressed with any of the Zn^{2+} binding GH mutants (including also double and triple mutants). Interestingly, extracellular secretion and intracellular production of the GH mutants expressed singly was about 50% lower when compared to that of *wt*-GH which suggests the role of Zn^{2+}-binding residues in GH stability having a final impact on secretion.

However, *wt*-GH and/or GH mutant having alanines at all three high-affinity Zn^{2+} binding residues transiently transfected in GC cells (rat pituitary cells that do not produce endogenous rat GH), displayed significantly reduced secretion (analyzed by DSL-ELISA) of the GH mutant compared to the *wt*-GH after forskolin stimulation [Petkovic and Mullis, unpubl. data]. At the same time, intracellular production of the mutant was not different from that of the *wt*-GH (analyzed by Western blot). These data indicate that loss of affinity of GH to Zn^{2+} could interfere with normal GH dimerization/aggregation as a prerequisite step towards normal secretion.

Similarly to PRL, aggregation of endogenous rat GH was investigated in GH_4C_1 cells using the pulse-chase method in combination with detection of protein aggregates via their insolubility in Lubrol [10]. The reported data revealed GH to be soluble in Lubrol immediately after synthesis while 30 min later, roughly 40–50% of newly synthesized hormone acquires the form of aggregate (Lubrol-insoluble form) with the same time course and to about the same maximum extent as PRL [10]. Moreover, 'labeled' human GH transiently transfected in AtT-20 cells became over 40% aggregated 30 min after synthesis in that way following the same kinetics as those for the rGH in GH_4C_1 cells. Furthermore, aggregation of GH seems to be specific for pituitary cells since in COS cells (fibroblast cells) that do not store secretory proteins in granules, it occurred far less efficiently (10%) than in AtT-20 cells using the same experimental conditions. The high concentration of Zn^{2+} present in the secretory pathway of pituitary cells might be one reason why GH aggregates in pituitary cell lines and not well in fibroblasts. As already mentioned earlier, pH is one of the specific factors that play a role in the process of protein aggregation. Neutralizing the pH of the *trans*-Golgi lumen in AtT-20 cells slowed aggregation of human GH to a lesser extent than it did to that of human PRL [10, 15], suggesting that aggregation of GH is possibly less dependent on acidic pH than that of PRL. These data also indicate that despite high structural similarity to PRL, the specific factors required for efficient aggregation of GH compared to PRL are not necessarily the same.

Zn^{2+} binding to GH through high-affinity binding sites is necessary for GH dimerization. Whether the dimers are the main storage form of GH in secretory granules or whether additional intramolecular cross-linking occurs through low-affinity Zn^{2+} binding with other amino acids enhancing in that way GH aggregation and storage in secretory granules, still remains to be elucidated.

Zinc Transporters and Their Role in the Early Secretory and the Regulated Secretory Pathway

Zn^{2+} is required as a structural and/or catalytic cofactor by hundreds of different proteins. Many zinc-dependent proteins reside in organelles of the secretory pathway or pass through this pathway on their way to other compartments (e.g. vacuole, lysosomes) within the cell or prior to their secretion into the extracellular environment. Zn^{2+} homeostasis is therefore tightly regulated and because of its high charge density, transporter proteins are required to move it across cellular membranes. Zinc transporters that mediate Zn^{2+} homeostasis and dynamics in the secretory pathway fulfil this crucial role. To date, more than 20 zinc transporters have been identified, characterized and classified into two families (Slc30a/ZnT and Slc39a/Zip), nearly half of which is localized to the early secretory and regulated secretory pathways (fig. 2) [27, 28]. Most ZnT transporters have six transmembrane domains (TMDs) and their function is to reduce intracellular Zn^{2+} availability by promoting zinc efflux into the extracellular space as well as transport of Zn^{2+} from the cytosol into the intracellular organelles. On the other hand, Zip transporters are predicted to have eight TMDs and to function in the opposite way to ZnT transporters, i.e. to increase intracellular Zn^{2+} availability by promoting extracellular Zn^{2+} uptake and, perhaps, vesicular Zn^{2+} release into the cytoplasm.

Of all the ZnTs localized to subcellular compartments throughout the secretory pathway, ZnT5, ZnT6 and ZnT7 have been identified to localize in the early secretory pathway (fig. 2) especially in the ER and Golgi complex as shown by zinc transport assay using radioactively labeled Zn^{2+} and by zinc-staining analysis using fluorescent zinc probes [29, 30]. Their role is to assure the transport of Zn^{2+} into the lumen of these organelles even though the results of these studies suggest that zinc transport is not robust. Interestingly, overexpression of ZnT5–7 does not provide significant resistance to high concentrations of Zn^{2+}, while the total cellular Zn^{2+} content of DT40 cells (chicken B lymphocyte-derived cell line) lacking the same transporters is not significantly changed when compared to that of wild-type cells [31]. Hence, their role would be more to load Zn^{2+} to zinc-requiring enzymes involved in the early secretory pathway, as for example alkaline phosphatases that are essential for bone mineralization [32]. Direct evidence to confirm this hypothesis came from studies using an in vitro system based on DT40 cells in which the activation of tissue-nonspecific alkaline phosphatase (TNAP) was investigated. Combined gene disruption/re-

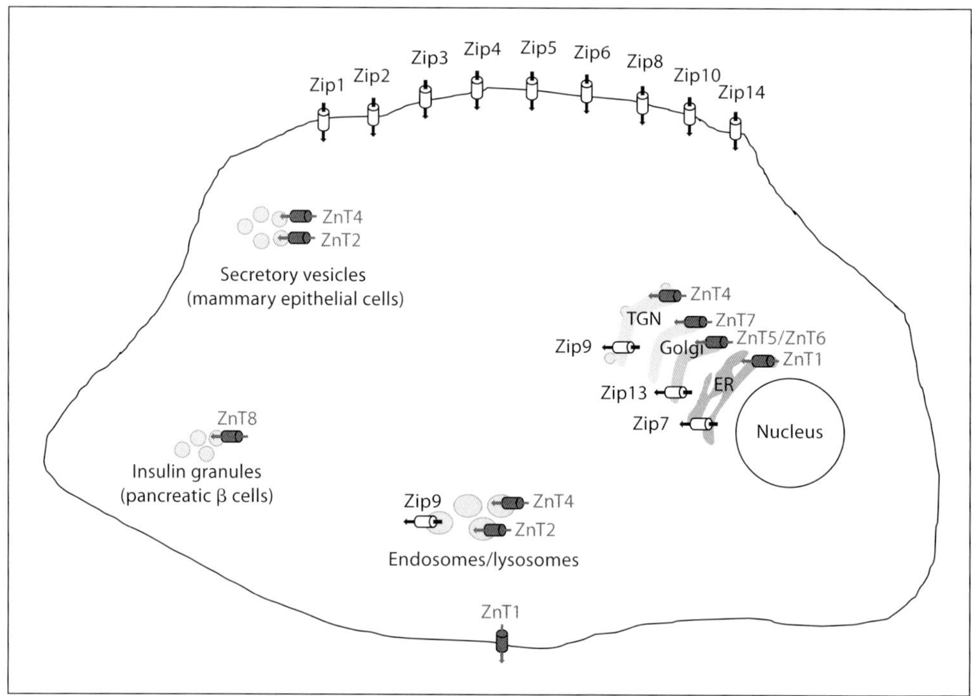

Fig. 2. Localization of zinc transporters and direction of zinc transport within the secretory pathway. Zip transporters (white cylinders) promote transport of extracellular Zn^{2+} into the cytoplasm, which is then taken up by ZnT transporters (dark grey cylinders) and released into the ER and Golgi complex (early secretory pathway) and into secretory vesicles/granules (regulated secretory pathway). The role of various Zip transporters is also to move Zn^{2+} out of the secretory pathway. The arrows represent direction of Zn^{2+} mobilized by Zip transporters (black arrow) or ZnT transporters (dark gray arrow), respectively.

expression experiments revealed that ZnT5–7 are all required for TNAP activation known to depend on Zn^{2+} [31, 33]. Deletion of ZnT5 gene in mouse leads to abnormal bone development, loss of weight and lethal, male-specific, cardiac arrhythmias [34]. Several studies suggest that the activation of TNAP in the early secretory pathway requires the formation of ZnT5/ZnT6 heterodimers while ZnT7 forms homo-oligomers (probably homodimers) [31, 35]. Another study indicates that ZnT5 is capable of independently mediating Zn^{2+} transport, but co-expression of ZnT5 and ZnT6 accelerates its rate [36]. In addition, the authors demonstrate that Zn^{2+} transport mediated by ZnT5 is coupled to changes in Golgi pH. While this links ZnT5/6 and vesicular Zn^{2+}, the precise underlying mechanism and regulation remains poorly understood. Moreover, ZnT5/ZnT6 heterodimers are the only complexes that function in most eukaryotes [37] and their cellular function in the homeostatic maintenance of early secretor pathway seems to be conserved evolutionary. The counterparts of ZnT5/ZnT6 heterodimers in yeast, Msc2p/Zrg17p (*Saccharomyces cerevisiae*) and

Cis4/ZRG17 (*Schizosaccharomyces pombe*), are required for proper functioning of the ER and for Golgi membrane trafficking [38, 39]. In addition, yeast strains lacking these complexes display a defect in ER-associated degradation and show increased unfolded protein response (UPR) [38]. As far as Zip transporters and their role in the early secretory pathway are concerned, they regulate the transport of Zn^{2+} in the opposite direction to ZnT and induce an increase of intracellular Zn^{2+} by promoting extracellular Zn^{2+} uptake (fig. 2). In addition, they may participate in fine tuning of the Zn^{2+} transport out of the ER and Golgi complex into the cytoplasm. In fact, Zip7, Zip9 and Zip13 have been reported to localize in the ER and Golgi complexes and to regulate Zn^{2+} content in the lumen of these organelles playing modulatory roles in the activity of zinc-requiring enzymes like TNAP [40–42].

The final step in the secretory pathway of endocrine, neuroendocrine and other protein-secreting cell types is to store proteins synthesized for secretion in secretory granules. As mentioned earlier, in many of these specialized cell types, like for example pancreatic β-cells, somatotrope cells of the anterior pituitary gland and secreting mammary epithelial cells, a high Zn^{2+} content has been detected [9, 43, 44]. Thus, Zn^{2+}-containing vesicles/granules must have specialized specific zinc transporters to regulate the supply of Zn^{2+} in and out of their lumen. During insulin biosynthesis Zn^{2+} plays a crucial role in insulin aggregation/crystalization (formation of the Zn^{2+}-proinsulin hexamers), which is required for its storage in insulin granules [45]. ZnT5 was the first zinc transporter reported to be associated with insulin granules in pancreatic β-cells and the formation of Zn^{2+}-insulin crystals [29]. Since in *ZnT5*-KO mice the formation of dense cores composed of Zn^{2+}-insulin crystals appears to be normal, ZnT5 was suggested to fulfil the role in zinc supply in the early secretory pathway, probably together with ZnT6. Furthermore, the discovery of ZnT8, as a ZnT specifically expressed pancreatic in β-cells, opened an entirely new field in the research of diabetes [46, 47]. A significant decrease of Zn^{2+} content analyzed in β-cells of *ZnT8*-KO mice together with loss of dense cores of Zn^{2+}-insulin crystals point towards essential role of ZnT8 in Zn^{2+}-insulin crystal formation [48, 49].

Considerably higher concentrations of Zn^{2+} found in breast milk than those found in serum suggest the existence of transport machinery in mammary epithelial cells responsible for the uptake and release of Zn^{2+} into the milk. Despite the fact that ZnT2 was first reported to facilitate Zn^{2+} transport into endosomes/lysosomes [50], the study of Chowanadisai et al. [51] revealed its importance in transporting Zn^{2+} into breast milk. They found a heterozygous mutation in the human *ZnT2* gene in women with low Zn^{2+} concentration in the milk leading to the diagnosis of severe Zn^{2+} deficiency in their breast-fed infants. A similar low Zn^{2+}-secreting phenotype was found in the lethal milk (lm) mouse with inherited Zn^{2+} deficiency caused by a mutation in the *ZnT4* gene [52].

Regarding the possible role of Zip transporters in the regulated secretory pathway, there have been no studies reported to date that describe their localization and/or function in secretory vesicles/granules.

Conclusions

One of the main hallmarks of endocrine and neuroendocrine cells is to store secretory proteins in concentrated forms in dense cores of secretory granules which are released upon appropriate stimulation leading to a burst of hormone in the bloodstream. During the process of secretory granule biogenesis, self-association (aggregation) of hormone destined for secretion seems to be one of the crucial steps. In the case of PRL and GH this process occurs in the presence of Zn^{2+}. Two Zn^{2+} are required for GH dimers to form and association occurs through binding at high-affinity residues in GH. It still remains to be elucidated whether GH aggregation could be enhanced by intramolecular cross-bridging of GH dimers that may occur through Zn^{2+} binding at low-affinity residues. Nevertheless, GH mutant at all three high-affinity Zn^{2+} binding residues displays significantly reduced secretion compared to the *wt*-GH when analyzed in GC cells. In addition, *wt*-GH secretion is significantly impaired when cells were cultured in Zn^{2+}-deficient versus complete culture medium [Petkovic and Mullis, unpubl. data]. Therefore, it is possible that loss of affinity of GH to Zn^{2+} as well as intracellular free (chelatable) Zn^{2+} content as such could interfere with normal GH storage in secretory granules and possibly explain the disturbed GH secretion.

The complexity and importance of cellular Zn^{2+} homeostasis is reflected by the large number and variety of proteins found in virtually every cell compartment that is potentially dedicated to Zn^{2+} transport, among them, the two families of zinc transporters. The localization and function of various ZnT and Zip transporters within the regulated secretory pathway has been well documented. It is becoming more evident that they might be responsible for the fine tuning of Zn^{2+} transport in and out of each of the organelles participating in the regulated secretory pathway (ER, Golgi complex and secretory granules). Thus, their role in supplying the secretory pathway with the sufficient amount of Zn^{2+} required for secretory granule biogenesis, which is crucial for normal secretion, may be greater than previously anticipated.

References

1 Anelli T, Sitia R: Protein quality control in the early secretory pathway. EMBO J 2008;27:315–327.

2 Losev E, Reinke CA, Jellen J, Strongin DE, Bevis BJ, Glick BS: Golgi maturation visualized in living yeast. Nature 2006;441:1002–1006.

3 Nakano A, Luini A: Passage through the Golgi. Curr Opin Cell Biol 2010;22:471–478.

4 Puthenveedu MA, Linstedt AD: Subcompartmentalizing the Golgi apparatus. Curr Opin Cell Biol 2005;17:369–375.

5 De Matteis MA, Luini A: Exiting the Golgi complex. Nat Rev Mol Cell Biol 2008;9:273–284.

6 Glombik MM, Gerdes HH: Signal-mediated sorting of neuropeptides and prohormones: secretory granule biogenesis revisited. Biochimie 2000;82:315–326.

7 Dannies PS: Mechanisms for storage of prolactin and growth hormone in secretory granules. Mol Genet Metab 2002;76:6–13.

8 Tooze SA, Martens GJ, Huttner WB: Secretory granule biogenesis: rafting to the SNARE. Trends Cell Biol 2001;11:116–122.

9 Thorlacius-Ussing O: Zinc in the anterior pituitary of rat: a histochemical and analytical work. Neuroendocrinology 1987;45:233–242.

10   Lee MS, Zhu YL, Chang JE, Dannies PS: Acquisition of Lubrol insolubility, a common step for growth hormone and prolactin in the secretory pathway of neuroendocrine cells. J Biol Chem 2001;276:715–721.

11   Giannattasio G, Zanini A, Meldolesi J: Molecular organization of rat prolactin granules. I. In vitro stability of intact and 'membraneless' granules. J Cell Biol 1975;64:246–251.

12   Deladoëy J, Stocker P, Mullis PE: Autosomal dominant GH deficiency due to an Arg183His GH-1 gene mutation: clinical and molecular evidence of impaired regulated GH secretion. J Clin Endocrinol Metab 2001;86:3941–3947.

13   Zhu YL, Conway-Campbell B, Waters MJ, Dannies PS: Prolonged retention after aggregation into secretory granules of human R183H-growth hormone (GH), a mutant that causes autosomal dominant GH deficiency type II. Endocrinology 2002;143:4243–4248.

14   Wu MM, Grabe M, Adams S, Tsien RY, Moore HP, Machen TE: Mechanisms of pH regulation in the regulated secretory pathway. J Biol Chem 2001;276:33027–33035.

15   Sankoorikal BJ, Zhu YL, Hodsdon ME, Lolis E, Dannies PS: Aggregation of human wild-type and H27A-prolactin in cells and in solution: roles of Zn(2+), Cu(2+), and pH. Endocrinology 2002;143:1302–1309.

16   Ellis RJ: Macromolecular crowding: obvious but underappreciated. Trends Biochem Sci 2001;26:597–604.

17   Minton AP: Implications of macromolecular crowding for protein assembly. Curr Opin Struct Biol 2000;10:34–39.

18   Cunningham BC, Mulkerrin MG, Wells JA: Dimerization of human growth hormone by zinc. Science 1991;253:545–548.

19   Sun Z, Li PS, Dannies PS, Lee JC: Properties of human prolactin (PRL) and H27A-PRL, a mutant that does not bind Zn++. Mol Endocrinol 1996;10:265–271.

20   Miura T, Suzuki K, Kohata N, Takeuchi H: Metal binding modes of Alzheimer's amyloid beta-peptide in insoluble aggregates and soluble complexes. Biochemistry 2000;39:7024–7031.

21   Berg JM, Shi Y: The galvanization of biology: a growing appreciation for the roles of zinc. Science 1996;271:1081–1085.

22   Coleman JE: Zinc enzymes. Curr Opin Chem Biol 1998;2:222–234.

23   Vallee BL, Falchuk KH: The biochemical basis of zinc physiology. Physiol Rev 1993;73:79–118.

24   Carroll RJ, Hammer RE, Chan SJ, Swift HH, Rubenstein AH, Steiner DF: A mutant human proinsulin is secreted from islets of Langerhans in increased amounts via an unregulated pathway. Proc Natl Acad Sci USA 1988;85:8943–8947.

25   Nicoll CS, Mayer GL, Russell SM: Structural features of prolactins and growth hormones that can be related to their biological properties. Endocr Rev 1986;7:169–203.

26   Iliev DI, Wittekindt NE, Ranke MB, Binder G: In vitro analysis of hGH secretion in the presence of mutations of amino acids involved in zinc binding. J Mol Endocrinol 2007;39:163–167.

27   Lichten LA, Cousins RJ: Mammalian zinc transporters: nutritional and physiologic regulation. Annu Rev Nutr 2009;29:153–176.

28   Sekler I, Sensi SL, Hershfinkel M, Silverman WF: Mechanism and regulation of cellular zinc transport. Mol Med 2007;13:337–343.

29   Kambe T, Narita H, Yamaguchi-Iwai Y, Hirose J, Amano T, Sugiura N, Sasaki R, Mori K, Iwanaga T, Nagao M: Cloning and characterization of a novel mammalian zinc transporter, zinc transporter 5, abundantly expressed in pancreatic beta cells. J Biol Chem 2002;277:19049–19055.

30   Kirschke CP, Huang L: ZnT7, a novel mammalian zinc transporter, accumulates zinc in the Golgi apparatus. J Biol Chem 2003;278:4096–4102.

31   Suzuki T, Ishihara K, Migaki H, Matsuura W, Kohda A, Okumura K, Nagao M, Yamaguchi-Iwai Y, Kambe T: Zinc transporters, ZnT5 and ZnT7, are required for the activation of alkaline phosphatases, zinc-requiring enzymes that are glycosylphosphatidylinositol-anchored to the cytoplasmic membrane. J Biol Chem 2005;280:637–643.

32   Zurutuza L, Muller F, Gibrat JF, Taillandier A, Simon-Bouy B, Serre JL, Mornet E: Correlations of genotype and phenotype in hypophosphatasia. Hum Mol Genet 1999;8:1039–1046.

33   Suzuki T, Ishihara K, Migaki H, Nagao M, Yamaguchi-Iwai Y, Kambe T: Two different zinc transport complexes of cation diffusion facilitator proteins localized in the secretory pathway operate to activate alkaline phosphatases in vertebrate cells. J Biol Chem 2005;280:30956–30962.

34   Inoue K, Matsuda K, Itoh M, Kawaguchi H, Tomoike H, Aoyagi T, Nagai R, Hori M, Nakamura Y, Tanaka T: Osteopenia and male-specific sudden cardiac death in mice lacking a zinc transporter gene, Znt5. Hum Mol Genet 2002;11:1775–1784.

35 Ishihara K, Yamazaki T, Ishida Y, Suzuki T, Oda K, Nagao M, Yamaguchi-Iwai Y, Kambe T: Zinc transport complexes contribute to the homeostatic maintenance of secretory pathway function in vertebrate cells. J Biol Chem 2006;281:17743–17750.

36 Ohana E, Hoch E, Keasar C, Kambe T, Yifrach O, Hershfinkel M, Sekler I: Identification of the Zn^{2+} binding site and mode of operation of a mammalian Zn^{2+} transporter. J Biol Chem 2009;284:17677–17686.

37 Kambe T, Suzuki T, Nagao M, Yamaguchi-Iwai Y: Sequence similarity and functional relationship among eukaryotic ZIP and CDF transporters. Genomics Proteomics Bioinformatics 2006;4:1–9.

38 Ellis CD, Macdiarmid CW, Eide DJ: Heteromeric protein complexes mediate zinc transport into the secretory pathway of eukaryotic cells. J Biol Chem 2005;280:28811–28818.

39 Fang Y, Sugiura R, Ma Y, Yada-Matsushima T, Umeno H, Kuno T: Cation diffusion facilitator Cis4 is implicated in Golgi membrane trafficking via regulating zinc homeostasis in fission yeast. Mol Biol Cell 2008;19:1295–1303.

40 Fukada T, Civic N, Furuichi T, Shimoda S, Mishima K, Higashiyama H, Idaira Y, Asada Y, Kitamura H, Yamasaki S, Hojyo S, Nakayama M, Ohara O, Koseki H, Dos Santos HG, Bonafe L, Ha-Vinh R, Zankl A, Unger S, Kraenzlin ME, Beckmann JS, Saito I, Rivolta C, Ikegawa S, Superti-Furga A, Hirano T: The zinc transporter SLC39A13/ZIP13 is required for connective tissue development; its involvement in BMP/TGF-beta signaling pathways. PLoS One 2008;3:e3642.

41 Huang L, Kirschke CP, Zhang Y, Yu YY: The ZIP7 gene (Slc39a7) encodes a zinc transporter involved in zinc homeostasis of the Golgi apparatus. J Biol Chem 2005;280:15456–15463.

42 Matsuura W, Yamazaki T, Yamaguchi-Iwai Y, Masuda S, Nagao M, Andrews GK, Kambe T: SLC39A9 (ZIP9) regulates zinc homeostasis in the secretory pathway: characterization of the ZIP subfamily I protein in vertebrate cells. Biosci Biotechnol Biochem 2009;73:1142–1148.

43 Lopez V, Kelleher SL: Zinc transporter-2 (ZnT2) variants are localized to distinct subcellular compartments and functionally transport zinc. Biochem J 2009;422:43–52.

44 Zalewski PD, Millard SH, Forbes IJ, Kapaniris O, Slavotinek A, Betts WH, Ward AD, Lincoln SF, Mahadevan I: Video image analysis of labile zinc in viable pancreatic islet cells using a specific fluorescent probe for zinc. J Histochem Cytochem 1994;42:877–884.

45 Dodson G, Steiner D: The role of assembly in insulin's biosynthesis. Curr Opin Struct Biol 1998;8:189–194.

46 Chimienti F, Devergnas S, Favier A, Seve M: Identification and cloning of a beta-cell-specific zinc transporter, ZnT-8, localized into insulin secretory granules. Diabetes 2004;53:2330–2337.

47 Chimienti F, Favier A, Seve M: ZnT-8, a pancreatic beta-cell-specific zinc transporter. Biometals 2005;18:313–317.

48 Lemaire K, Ravier MA, Schraenen A, Creemers JW, Van de Plas R, Granvik M, Van Lommel L, Waelkens E, Chimienti F, Rutter GA, Gilon P, in't Veld PA, Schuit FC: Insulin crystallization depends on zinc transporter ZnT8 expression, but is not required for normal glucose homeostasis in mice. Proc Natl Acad Sci USA 2009;106:14872–14877.

49 Wijesekara N, Dai FF, Hardy AB, Giglou PR, Bhattacharjee A, Koshkin V, Chimienti F, Gaisano HY, Rutter GA, Wheeler MB: Beta cell-specific Znt8 deletion in mice causes marked defects in insulin processing, crystallisation and secretion. Diabetologia 2010;53:1656–1668.

50 Palmiter RD, Cole TB, Findley SD: ZnT-2, a mammalian protein that confers resistance to zinc by facilitating vesicular sequestration. EMBO J 1996;15:1784–1791.

51 Chowanadisai W, Lonnerdal B, Kelleher SL: Identification of a mutation in SLC30A2 (ZnT-2) in women with low milk zinc concentration that results in transient neonatal zinc deficiency. J Biol Chem 2006;281:39699–39707.

52 Huang L, Gitschier J: A novel gene involved in zinc transport is deficient in the lethal milk mouse. Nat Genet 1997;17:292–297.

Dr. Vibor Petkovic, PhD
University Children's Hospital, Pediatric Endocrinology, Inselspital
CH–3010 Bern (Switzerland)
Tel. +41 31 632 0339, E-Mail vibor.petkovic@dkf.unibe.ch

Mullis P-E (ed): Developmental Biology of GH Secretion, Growth and Treatment.
Endocr Dev. Basel, Karger, 2012, vol 23, pp 109–120 (DOI: 10.1159/000341766)

Isolated Growth Hormone Deficiency Type 2: From Gene to Therapy

Maria Consolata Miletta · Didier Lochmatter · Vibor Pektovic · Primus-E. Mullis

University Children's Hospital, Pediatric Endocrinology, Diabetology and Metabolism, Inselspital, Bern, Switzerland

Abstract

Isolated growth hormone deficiency type-2 (IGHD-2), the autosomal-dominant form of GH deficiency, is mainly caused by specific splicing mutations in the human growth hormone (hGH) gene (*GH-1*). These mutations, occurring in and around exon 3, cause complete exon 3 skipping and produce a dominant-negative 17.5 kD GH isoform that reduces the accumulation and secretion of wild type-GH (*wt*-GH). At present, patients suffering from IGHD-2 are treated with daily injections of recombinant human GH (rhGH) in order to reach normal height. However, this type of replacement therapy, although effective in terms of growth, does not prevent toxic effects of the 17.5-kD mutant on the pituitary gland, which can eventually lead to other hormonal deficiencies. Considering a well-known correlation between the clinical severity observed in IGHD-2 patients and the increased expression of the 17.5-kD isoform, therapies that specifically target this isoform may be useful in patients with *GH-1* splicing defects. This chapter focuses on molecular strategies that could represent future directions for IGHD-2 treatment.

Introduction

GHRH-GH-IGF-1 Secretory Axis

Growth hormone (GH) is a member of the somatotropin/prolactin family of hormones and plays an important role in human physiology. It controls many physiological processes such as growth, bone mineralization, sugar and lipid metabolism, protein synthesis and stimulation of the immune system, and it is in turn influenced by important external factors such as stress, sleep, exercise and food intake [1–5]. GH is secreted from the somatotroph cells in the anterior pituitary gland and it is released in the bloodstream under regulation of two hypothalamic peptides: GH-releasing hormone (GHRH), which is stimulatory, and GH-inhibiting factor (GHIF; somatostatin), which is inhibitory [6]. Following secretion, GH is delivered to the liver and other target organs where it binds to the GH receptor (GHR) and mediates the release of insulin-like growth factor-1 (IGF-1), the primary mediator for the growth-mediating actions of GH [7]. IGF-1, in turn, is conjugated in the plasma to IGF-binding protein-3 (IGF-BP-3) [8] and to acid-labile subunit (ALS) [9], thus prolonging its half-life and modulating its availability to target tissues. It has, in fact, growth stimulating effects on a wide variety of tissues but its principally stimulatory effect is to promote growth development during childhood and puberty acting on osteoblast and chondrocyte. Further, GH normally regulates its own expression and production by both direct and indirect feedback (via IGF-1): IGF-1 leads to decreased secretion of GH by suppressing the somatotrophs and by stimulating release of somatostatin from the hypothalamus. In parallel, GH inhibits GHRH secretion and probably has a direct autocrine inhibitory effect on secretion from the somatotrophs. Integration of all the factors affecting GH synthesis and secretion lead to a pulsatile pattern of release.

GH-1 Gene Cluster

The *GH-1* gene consists of five exons and four introns, enabling transcription of five known variants through alternative splicing. When *GH-1* is correctly spliced, it produces the 22-kD isoform, which includes all five exons and represents the major biologically active form of hGH present in the circulation. X-ray crystallographic studies have shown that it comprises a core of two pairs of parallel α-helices arranged in an up-up-down-down fashion and stabilized by two intramolecular disulphide linkages (Cys53–Cys165, Cys182–Cys189) [10]. However, even under normal condition, a small percentage of alternative splicing products is generated. These splicing variants include: (a) the 20-kD isoform (representing 5–10% of GH transcripts), as a result of the activation of an in-frame cryptic splice site within exon 3 which causes deletion of amino acids 32–46. From the structural point of view, the 20-kD isoform lacks part of the loop connecting the first two helices of the GH molecule [11, 12]; (b) the 17.5-kD isoform, another GH splicing variant (representing 1–5% of GH transcripts) resulting from complete skipping of exon 3, which is the entire loop that connects helix 1 and 2 in the tertiary structure of GH generating truncated GH isoform with no biological activity; (c) the severely truncated 11.3- and 7.4-kD isoforms, present in trace amount in normal human pituitary and encoded by transcripts that skip exons 3–4 or exons 2–4 yielding inactive protein products [13]. Multiple mechanisms have evolved to avoid production of skipped *GH-1* transcripts, especially those encoding the 17.5-kD isoform [14]. Since *GH-1* has several weak splice sites and exhibits small amounts of aberrant splicing even in healthy individuals, multiple enhancers (*cis*-acting splicing regulatory elements) are essential to maintain full-length GH splicing. They contribute to exon 3 definition and to the enhancer-dependent activation of the intron 2,3 splice site at the expense of the stronger, nearby cryptic splice site [15]. Up to now, three *GH-1* splicing enhancers are well characterized: ESE1, exon splice enhancer motif 1, which encompasses the first seven bases of exon 3; ESE2, a second exon splicing enhancer, located 12 nt upstream of the cryptic splice site in exon 3; and

ISE, intron splicing enhancer contained in intron 3. ESE1 strengthens the use of the weak 3′ splice site upstream of exon 3 suppressing at the same time, a downstream cryptic splice site. Mutations in any of the bases of ESE1 lead to either complete or partial exon 3 skipping and generation of the 17.5- and 20-kDa isoforms at various amounts [16, 17]. ESE2 has been reported to contribute the most to the splicing activation of all three splicing enhancers, suggesting that it might also regulate the majority of *GH-1* splicing in cells. This is supported by the observation that point mutations within the ESE2 cause more aberrant splicing than point mutations in either ESE1 or ISE. Point mutations which disrupt ISE1 exhibit exon 3 skipping but in a low percentage. Besides enhancers, the overall size of intron 3 seems also to be crucial for exon inclusion. Intron 3 in *GH-1* is 92 nt in length, well above the approximately 50 nucleotide minimum intron length required for spliceosome assembly. Decreasing the size of intron 3, rather than deleting the specific sequences within intron 3, by as little as 12 nt, prevents its accurate identification by the spliceosome causing increased exon 3 skipping [14].

IGHD-2

Isolated growth hormone deficiency type-2 (IGHD-2) is an autosomal-dominant disorder caused by mutations in *GH-1* gene [13]. In contrast to nonsense, missense, and deletion mutations that characterize autosomal-recessive growth hormone deficiency (IGHD-1), the majority of IGHD-2 mutations described so far affect the splicing of the *GH-1* gene. In the majority of cases, IGHD-2 mutations occur within the first 6 bp of intervening sequences 3 (5′ IVS-3) or within ESE and ISE regions [18, 19] and result in skipping of exon 3 during the process of alternative splicing leading to overproduction of the 17.5-kD GH isoform. GH mutations disturbing conserved consensus splice sites bordering exon 3 cause almost complete skipping of exon 3, by impairing U1 small nuclear ribonucleoprotein (snRNP) recognition and spliceosome assembly, while mutations disrupting splicing enhancers lead to different amounts of exon 3-skipped GH isoform. For example, subjects presenting with a 5′ IVS-3 +1/+2 bp splice site mutation show a more severe IGHD-2 phenotype when compared with subject presenting with a 5′ IVS-3 +5/+6 bp, probably due to a larger quantity of 17.5-kD GH isoform produced [20]. In addition to splice-site mutations, GH missense mutations (P89L, R183H, V110F and R178H) have also been reported to cause IGHD-2 [21, 22].

The loss of exon 3 in GH molecule deletes the linker domain between the first two helices of GH, disrupting an internal disulfide bridge, and thereby affecting overall protein structure. The 17.5-kD isoform generated has been reported to exhibit a dominant-negative effect on secretion of the *wt*-GH in cell culture and transgenic mice [16, 23]. This isoform is initially retained in the endoplasmic reticulum (ER) where it triggers the unfolded protein response disrupting the secretory pathway and trafficking of GH and other hormones, including adrenocorticotropic hormone (ACTH). After being produced, the 17.5-kD isoform is detected by ER quality control mechanism and degraded via the proteasomal pathway. When the production rate of 17.5-kD isoform exceeds the proteasome degradation capacity, it accumulates in the cytoplasm leading to reduced cell proliferation and apoptosis of GC cells [24]. The situation is more complex in vivo because in IGHD-2 patients the GH-negative feedback is reduced and it results in a chronic upregulation of GHRH and reduction of somatostatin expression leading to an increased stimulatory drive to GH expression, both from the normal and mutant GH alleles [25]. In mice expressing a high copy number GH-IVS+1 mutant transgene (producing exclusively the 17.5-kD hGH isoform) mouse GH was greatly reduced. This reduction was accompanied by the disruption of the somatotroph secretory pathway and mice were presenting a dwarf phenotype [25]. Moreover, large numbers of infiltrating macrophages were observed in the pituitaries of these mutant animals, which showed almost total loss of somatotrophs and marked anterior pituitary gland hypoplasia associated with deficits in other pituitary-derived hormones [15, 25]. This secondary phenotype was copy number dependent: prolactin and TSH (and LH in males) were significantly reduced in adult high-copy animals, whereas these deficits were milder in low-copy transgenics. Furthermore, the onset and severity of IGHD-2 were also transgene copy number dependent; high-copy lines rapidly developed profound GHD by weaning, whereas low-copy mice showed milder adult-onset of GHD.

Clinical Implications: Variable and Evolving IGHD-2 Phenotype in Patients
Patients with IGHD-2 show wide variability in the age of onset and in the severity of IGHD-2. Clinical features that describe this disorder include low but detectable serum GH levels combined with pathologically low IGF-1 concentration leading to short stature and in severe cases, anterior pituitary gland hypoplasia. Reported pedigrees with Arg183His or E3 + 1 G→A mutations highlight the fact that patients with the same mutation can vary considerably in height and even attain normal adult height without treatment [26]. Moreover, splice-site mutations which exclusively produce the 17.5-kD isoform cause, on average, an earlier age of onset and greater clinical severity [27], while ESE mutations cause a slightly less severe impact on patients height [16]. The quantitative and qualitative differences in phenotype may result from variable splice-site mutations causing different amounts of 17.5-kD isoform and may reflect a threshold and a dose dependency effect of the amount of 17.5-kD relative to 22-kD GH isoform produced [25, 28]. The lack of negative feedback on GHRH in IGHD-2 patients causes constant upregulation of GHRH and leads to increase of the stimulatory drive on somatotrophs. This creates a vicious circle increasing the expression of the 17.5-kD isoform that further accelerates damage of the anterior pituitary leading to the development of other hormonal deficiencies. Thus, this model could explain why GHD patients develop combined pituitary hormone deficiency over time [24, 29]. This evolving phenotype is unpredictable and suggests the need for lifelong follow-up and possibly rhGH treatment in affected individuals. However, this type of approach is highly costly and an alternative therapeutic strategy would be more than welcome.

 Miletta · Lochmatter · Pektovic · Mullis

Supplementation of exogenous GH therapy with somatostatin could be an option: it could increase the efficacy of GH therapy, since in patients with splicing mutations this could effectively inhibit production of endogenous GH, thereby preventing production of the 17.5-kD isoform to progressively damage the remaining pituitary cells. Moreover, considering that normal GH secretion depends, at least in part, on avoiding increasing levels of the 17.5-kD isoform and its deleterious effects, novel therapies that specifically target this GH isoform may be useful to treat IGHD-2 patients with splicing mutations while avoiding the side-effects of long-term post-pubertal exogenous GH therapy.

IGHD-2: Therapeutic Approaches by RNA Interference

Rescue of IGHD-2 in vitro and in vivo by RNAi
In the last decade, few areas of biology have been transformed as thoroughly as RNA molecular biology. One of the most significant advances has been the discovery of small (20–30 nucleotide) noncoding RNAs that regulate genes and genomes. The effects of small RNAs on the control of gene expression are generally inhibitory, and the corresponding regulatory mechanisms are therefore collectively subsumed under the heading of RNA silencing and/or RNA interference (RNAi). Two primary categories of these small RNAs, short interfering RNAs (siRNAs) and microRNAs (mi-RNAs) act in both somatic and germline lineages of eukaryotic species to regulate endogenous genes and to defend the genome from invasive nucleic acids. Recent advances have revealed unexpected diversity in their biogenesis pathways and the regulatory mechanisms that they access [30]. RNAi has many potential advantages over traditional therapies, including increased specificity and versatility. Many recent advances in RNAi therapeutics have been used to successfully knock down viral genes or disease alleles [14, 31]. In 2004, Ryther et al. [14] proposed a strategy to use siRNAs specifically targeting the 17.5-kD GH isoform mRNA. They targeted the exon 3 skipped transcripts for degradation by designing siRNA for the exon 2-exon 4 junction, because it represents the only sequence within the *GH-1* gene that is unique to transcripts encoding the 17.5-kD isoform. In cultured GH3 cells, plasmids expressing this siRNA efficiently and specifically degraded the exon 3-skipped GH transcripts by almost 90%. These data suggested that siRNAs could be designed to target dominant, aberrantly spliced isoforms or other dominant alleles for degradation. Soon thereafter, Shariat et al. [32] were able to rescue pituitary function in IGHD-2 mice progeny by mating them with transgenic mice expressing a short hairpin RNAs (shRNAs) targeting the 17.5-kDa isoform. This study demonstrated the efficacy of siRNA or shRNA against the 17.5-kDa isoform. Clearly this methodology presents some limits: germline delivery of this type is impossible in the treatment of IGHD-2 patients, and using siRNAs and shRNAs may induce cytotoxicity in vitro and in vivo [33, 34]. A possible alternative could be to develop an RNA silencing approach in a more relevant direct gene therapy context for correction of somatic mutations and to use miRNAs that are

less prone to interfere with the endogenous RNAi pathway [35, 36]. Moreover, the difficulty to transfect certain cell types and the need of permanent gene knockdown can be overcome by using retroviral vectors (RVs) and lentiviral vector (LVs) enabling RNAi. LVs, especially, are widely used as promising vehicles for gene therapy because they have the ability to introduce their DNA into the genome of infected cells [35]. They can infect dividing and nondividing cell types unlike other RVs that require cell division for infection due to the fact that they cannot enter the nuclear envelope of a nondividing cell [37]. These vectors are being increasingly used as they are easy to handle and are able to transduce a large variety of cell types. In addition, constant development is being made to improve their safety, production or target specificity, making them an appreciated vehicle for gene therapy. Several strategies to deliver genes to specific tissues to heal disease such as spinal muscular atrophy (SMA) [38], hemophilia B [39], or SCID [40] are under development, but so far there has been no success in developing an approach to target GHD related disease. Lochmatter's group developed and tested a strategy involving LVs expressing a human micro-RNA-30-based shRNA (shRNAmir) [41, 42], targeting the anomalous junction of exon 2 and exon 4 (shRNAmir-Δ3) of the *GH-1* gene present only in the misspliced 17.5-kD isoform mRNA. Using rat pituitary tumor GC cells expressing both human *wt*-GH and Δ3GH, they demonstrated that shRNAmir efficiently reduces the expression of mutant protein, leading to an increase of *wt*-GH secretion, without affecting the viability of the transduced cells. Furthermore, cells expressing Δ3GH upon doxycycline (DOX) induction transduced with shRNAmir-Δ3 significantly reduced the 17.5-kD GH isoform at protein level and improved human *wt*-GH secretion in comparison to an shRNAmir targeting a scrambled sequence. Importantly, no toxicity due to shRNAmir expression could be observed in cell proliferation assays. Furthermore, confocal microscopy data strongly suggested that shRNAmir-Δ3 enabled the recovery of GH granule storage and secretory capacity. Taken together, these viral vectors have shown their ability to stably integrate, express shRNAmir and rescue IGHD type 2 phenotype in rat pituitary cells. This methodology might open new perspectives for the development of gene therapy to treat IGHD-2 patients. Nevertheless, several issues remain unresolved and must be carefully evaluated before the methodology can be considered a contender for therapy [31]. Even though the efficacy of siRNA against the 17.5-kD isoform has been shown, its delivery to specific tissue in vivo is still difficult to achieve. Clinical use of siRNAs will require an effective and safe delivery system, one of many obstacles to their potential therapeutic use. However, IGHD-2 may be an ideal system in which to study the use of therapeutic siRNA because the pituitary is not protected by the blood-brain barrier and may be reasonably accessible through stereotaxic surgery than blood supply. In addition, delivery of even small amounts of the siRNA coding the 17.5-kD (17.5-kD-si) isoform may prove efficacious because it seems to be very effective at low concentrations as suggested by Ryther and coworkers [15, 32]. Thus, delivery of even a small amount of siRNAs to the pituitary might possibly restore normal GH secretion in IGHD-2 patients.

Rescue of IGHD-2 by Pharmacologic Correction of Aberrant Splicing

Splicing Regulation as a Potential Genetic Modifier

Approximately 15% of all mutations that cause genetic diseases affect splicing of pre-mRNA [43]. Generally, pre-mRNA splicing involves precise removal of introns from pre-mRNA in a way that exons are spliced together to form mature mRNA with intact translation reading frames. Splicing requires exon recognition, followed by accurate cleavage and rejoining, which are determined by the invariant GT and AG intronic dinucleotides at the 5′ (donor) and 3′ (acceptor) exon-intron junctions. However, these invariant short-sequence motifs are not the exclusive determinants of accurate and efficient splicing. Several examples of intronic and exonic *cis*-elements that are important for correct splice-site identification and that are distinct from the classical splicing signals have been described. ESE and ISE promote and stimulate exon recognition while exonic and intronic silencers (ESS and ISS, respectively) inhibit recognition of splice sites and repress splicing. Both seem to be especially relevant for the regulation of alternative splicing [44]: ESEs in particular, appear to be very prevalent, and might be present in most exons, if not all, including constitutive ones [45, 46]. They are thought to serve as binding sites for specific serine/arginine-rich (SR) proteins [47, 48]. These proteins are part of a growing family of structurally related and highly conserved splicing factors characterized by the presence of 1–2 RNA recognition motifs (RRM) and by a different carboxy-terminal domain that is highly enriched in Arg/Ser dipeptides (the RS domain). The RRMs mediate sequence-specific binding to RNA, determining in that way substrate specificity, whereas the RS domain seems to be involved mainly in protein–protein interactions. SR proteins that are bound to ESEs can promote exon definition by directly recruiting the splicing machinery through their RS domain and/or by antagonizing the action of nearby silencer elements. These two models of splicing enhancement are not necessarily mutually exclusive, as they might reflect different requirements in the context of different exons. Most importantly, SR proteins modulate selection of alternative splicing sites in a concentration-dependent and tissue-specific manner. Overexpression of splicing factors regulate the splicing pattern of alternatively spliced exons and differences in the levels of alternatively spliced transcripts were suggested to arise from differences in the relative levels of splicing factors (the SR and hnRNP families) [48]. The possibility that altering the regulation of alternative splicing might restore normal phenotype in inherited diseases caused by aberrant splicing, led to the development of molecular approaches aiming to increase the level of correctly spliced transcripts through overexpression of splicing factors involved. Several small molecules, classified as histone deacytalase inhibitors (HDACi), such as sodium butyrate (NaB), valproic acid (VPA), 4-phenylbutyrate, SAHA, M34491–96 have been recently identified and reported to fulfill this aim [49, 50]. HDACi are a relatively new group of epigenetic agents that have multiple substrates including histone and nonhistone proteins, suggesting their involvement in multiple cellular processes. They alter the acetylation status of chromatin and other nonhistone proteins, resulting in changes in gene expression. In normal cells (noncancer cells), the response to HDACi depends on the cell type, the structure and concentration of HDACi as well as the exposure time to HDAC inhibitors [49]. The most promising results, using this molecular approach, come from in vitro and in vivo studies for treatment of cystic fibrosis (CF) and spinal muscular atrophy (SMA) [51–53]. Spinal muscular atrophy (SMA) is an autosomal-recessive neuromuscular disease characterized by degeneration of alpha motor neurons in the spinal cord. It is caused by homozygous mutations of the survival motor neuron 1 (SMN1) gene. Normally, both gene copies (*SMN1* and *SMN2*) are expressed, but they differ in the expression of full-length protein. *SMN2* gene preferentially gives rise to a truncated and less stable version of the SMN protein and thus cannot compensate for SMN1 loss or mutations unless it is not present in multiple copies. The differences between these highly homologous genes are in their RNA expression patterns. Most *SMN2* transcripts lack exons 3, 5, or most frequently, 7, with only a small amount of full-length mRNA generated. On the other hand, the *SMN1* gene expresses mostly a full-length mRNA, and only a small fraction of its transcripts are spliced to remove exons 3, 5 or 7. The amount of exon 7-containing SMN protein has been shown to be an inverse indicator of disease severity in SMA patients and mice. Therefore, increasing the expression of intact SMN protein may have clinically therapeutic effects on SMA patients. Chang et al. [51] showed that NaB treatment of human SMA lymphoid cell lines increased the expression of exon 7-containing SMN protein from the *SMN2* gene. After NaB treatment in vitro and in vivo, the transcription pattern of *SMN2* changed to an *SMN1*-like transcription pattern, which was almost identical to the *SMN* pattern in healthy individuals. Moreover, the clinical symptoms of the SMA-like mice improved after treatment with NaB.

Following this initial report, Nissim-Rafinia et al. [52] restored the function of the CFTR (cystic fibrosis transmembrane conductance regulator) channel: the ABC transporter class ion channel that transports chloride and thiocyanate across epithelial cell membranes. Mutations of the CFTR gene affect functioning of the chloride ion channels in epithelial cell membranes, leading to cystic fibrosis. The CFTR gene comprises 27 constitutively spliced exons, however several exons (3, 4, 9, 12, 14a, 16, 17b and 22) undergo in some individuals, partial aberrant splicing leading to a decrease in the level of full-length CFTR transcripts generating nonfunctional CFTR proteins. NaB treatment of cystic fibrosis-derived epithelial cells carrying the $3,849 + 10\,kb\;C{\rightarrow}T$ mutation increased the level of normal and full-length CFTR transcripts, restore the CFTR function which was sufficient to achieve channel activation. In both studies, NaB restores the correct splicing of pre-mRNA by overexpression of SR protein and transcription of both genes is modified through the alternative splicing rather than directly through inhibition of histone deacetylation. Even though discovery of chemical substances that have the potential ability to modulate alternative splicing seems promising, many of the compounds identified so far are not suitable for long-term therapy because of their unfavorable pharmacologic properties including toxicity and other undesirable side effects. In contrast, some HDACi such as VPA and 4-phenylbutyrate have already been approved by the Food and Drug Administration for application in the therapy of various diseases. VPA has successfully been used in the long-term therapy of epilepsy as well as for the treatment of mood disorders and migraine. The drug has clinically well-known and desirable pharmacologic properties including a suitable terminal half-life of 9–18 h in human serum. Certainly the most safe among HDACi is NaB and its derivatives. In fact, for several years, NaB has been used clinically to treat patients with sickle cell anemia and talassemia. The pharmacokinetics and toxicity studies of this molecule demonstrated that it has low toxicity and that it is well tolerated in both human and animal studies. A triglyceride analogue of butyric acid (tributyrin), which has 95% similarity in structure with NaB, has already been approved as a food additive in the US and it was shown that in clinics, after oral administration, it is well tolerated even by young children. Moreover, because tributyrin contains three butyric acid moieties esterified to glycerol, when completely hydrolyzed by cellular lipases or esterases, it yields 3-fold more butyric acid than NaB, producing and maintaining higher serum butyrate levels with prolonged half-life [54]. Pilot trials with phenylbutyrate in SMA patients already revealed promising results. However, only phase 3 placebo-controlled clinical trials will provide the final proof for a benefit of HDACi in therapy. Furthermore, several critical issues about HDACi like the precise action mechanisms, their influence on cell signaling pathways, long- and short-term consequences on the molecular profile of patients, and the use of different doses and routes of administration in combination treatments still remain to be elucidated.

IGHD-2, as SMA and CF, can be considered primarily as a splicing disorder in which severity of disease correlates with the level of correctly spliced RNA and with the ratio of alternatively spliced isoform. Therefore, a role for splicing regulation as a genetic modifier has been suggested. These recent studies provide direct evidence that changes in the level of splicing factors can modulate the splicing pattern of disease-associated genes and by increasing the level of a full-length transcript can restore protein function. Thus, this methodology could provide a useful approach for the treatment of IGHD-2. In fact, preliminary data from our lab indicate that NaB treatment significantly increases extracellular GH secretion of *wt*-GH and/or different GH-splice site mutants (IVS+2, IVS+6 and ISE+28) in GHRH-stimulated and nonstimulated conditions after transient transfection in rat pituitary cell line stably expressing hGHRHR (GC-GHRHR). Moreover, NaB decreased 17.5-kD/22-kD GH ratio, the key determinant of IGHD-2, both at transcriptional and at protein level [Miletta and Mullis, unpubl. data]. Hence, this could potentially open a new avenue of therapeutic interventions in IGHD-2 and other dominant disorders.

Conclusions

The expanding knowledge on the molecular defects leading to isolated growth hormone deficiency and the notion that the endocrine phenotype can evolve, are a challenge to traditional treatments. In future, a reclassification of the types of growth hormone deficiency may be made on the basis of genetic mechanism in combination with the phenotypic characteristic. Molecular approaches to therapy can target the pre-mRNA or the splicing factor will help to optimize the treatment of isolated growth hormone deficiency. Clear progress in research targeting alternative splicing has also been rapid and the field is quickly moving closer to clinical application.

References

1 Norrelund H: The metabolic role of growth hormone in humans with particular reference to fasting. Growth Horm IGF Res 2005;15:95–122.
2 Meinhardt UJ, Ho KK: Modulation of growth hormone action by sex steroids. Clin Endocrinol (Oxf) 2006;65:413–422.
3 Van Cauter E, Latta F, Nedeltcheva A, Spiegel K, Leproult R, Vandenbril C, Weiss R, Mockel J, Legros JJ, Copinschi G: Reciprocal interactions between the GH axis and sleep. Growth Horm IGF Res 2004;14(suppl A):S10–S17.
4 Widdowson WM, Healy ML, Sonksen PH, Gibney J: The physiology of growth hormone and sport. Growth Horm IGF Res 2009;19:308–319.
5 Walenkamp MJ, Wit JM: Genetic disorders in the GH IGF-1 axis in mouse and man. Eur J Endocrinol 2007;157(suppl 1):S15–S26.
6 Bluet-Pajot MT, Epelbaum J, Gourdji D, Hammond C, Kordon C: Hypothalamic and hypophyseal regulation of growth hormone secretion. Cell Mol Neurobiol 1998;18:101–123.
7 Liu JL, Le Roith D: Insulin-like growth factor I is essential for postnatal growth in response to growth hormone. Endocrinology 1999;140:5178–5184.
8 Hwa V, Oh Y, Rosenfeld RG: The insulin-like growth factor-binding protein (IGFBP) superfamily. Endocr Rev 1999;20:761–787.
9 Baxter RC: Circulating levels and molecular distribution of the acid-labile (alpha) subunit of the high molecular weight insulin-like growth factor-binding protein complex. J Clin Endocrinol Metab 1990;70:1347–1353.
10 De Vos AM, Ultsch M, Kossiakoff AA: Human growth hormone and extracellular domain of its receptor: crystal structure of the complex. Science 1992;255:306–312.
11 McCarthy EM, Phillips JA 3rd: Characterization of an intron splice enhancer that regulates alternative splicing of human GH pre-mRNA. Hum Mol Genet 1998;7:1491–1496.
12 Wada M, Ikeda M, Takahashi K, Asada N, Chang KT, Takahashi M, Honjo M: The full agonistic effect of recombinant 20kDa human growth hormone (hGH) on CHO cells stably transfected with hGH receptor cDNA. Mol Cell Endocrinol 1997;133:99–107.
13 Procter AM, Phillips JA, Cooper DN: The molecular genetics of growth hormone deficiency. Hum Genet 1998;103:255–272.
14 Ryther RC, Flynt AS, Harris BD, Phillips JA 3rd, Patton JG: GH1 splicing is regulated by multiple enhancers whose mutation produces a dominant-negative GH isoform that can be degraded by allele-specific small interfering RNA (siRNA). Endocrinology 2004;145:2988–2996.
15 Ryther RC, McGuinness LM, Phillips JA 3rd, Moseley CT, Magoulas CB, Robinson IC, Patton JG: Disruption of exon definition produces a dominant-negative growth hormone isoform that causes somatotroph death and IGHD-2. Hum Genet 2003;113:140–148.
16 Moseley CT, Mullis PE, Prince MA, Phillips JA 3rd: An exon splice enhancer mutation causes autosomal dominant GH deficiency. J Clin Endocrinol Metab 2002;87:847–852.
17 Petkovic V, Lochmatter D, Turton J, Clayton PE, Trainer PJ, Dattani MT, Eble A, Robinson IC, Fluck CE, Mullis PE: Exon splice enhancer mutation (GH-E32A) causes autosomal dominant growth hormone deficiency. J Clin Endocrinol Metab 2007;92:4427–4435.
18 Cogan JD, Prince MA, Lekhakula S, Bundey S, Futrakul A, McCarthy EM, Phillips JA 3rd: A novel mechanism of aberrant pre-mRNA splicing in humans. Hum Mol Genet 1997;6:909–912.
19 Takahashi I, Takahashi T, Komatsu M, Sato T, Takada G: An exonic mutation of the GH-1 gene causing familial isolated growth hormone deficiency type 2. Clin Genet 2002;61:222–225.

20 Mullis PE, Robinson IC, Salemi S, Eble A, Besson A, Vuissoz JM, Deladoey J, Simon D, Czernichow P, Binder G: Isolated autosomal dominant growth hormone deficiency: an evolving pituitary deficit? A multicenter follow-up study. J Clin Endocrinol Metab 2005;90:2089–2096.

21 Duquesnoy P SD, Netchine I, Dastot F, Sobrier ML, Goosens M: Familial isolated growth hormone deficiency with slight height reduction due to a heterozygote mutation in GH gene. Program 80th Ann Meet Endocrine Society, New Orleans, 1998, pp P2–P202.

22 Zhu YL, Conway-Campbell B, Waters MJ, Dannies PS: Prolonged retention after aggregation into secretory granules of human R183H-growth hormone (GH), a mutant that causes autosomal dominant GH deficiency type 2. Endocrinology 2002;143:4243–4248.

23 Cogan JD, Phillips JA 3rd, Schenkman SS, Milner RD, Sakati N: Familial growth hormone deficiency: a model of dominant and recessive mutations affecting a monomeric protein. J Clin Endocrinol Metab 1994;79:1261–1265.

24 Petkovic V, Godi M, Lochmatter D, Eble A, Fluck CE, Robinson IC, Mullis PE: Growth hormone (GH)-releasing hormone increases the expression of the dominant-negative GH isoform in cases of isolated GH deficiency due to GH splice-site mutations. Endocrinology 2010;151:2650–2658.

25 McGuinness L, Magoulas C, Sesay AK, Mathers K, Carmignac D, Manneville JB, Christian H, Phillips JA 3rd, Robinson IC: Autosomal dominant growth hormone deficiency disrupts secretory vesicles in vitro and in vivo in transgenic mice. Endocrinology 2003;144:720–731.

26 Binder G, Nagel BH, Ranke MB, Mullis PE: Isolated GH deficiency (IGHD) type 2: imaging of the pituitary gland by magnetic resonance reveals characteristic differences in comparison with severe IGHD of unknown origin. Eur J Endocrinol 2002;147:755–760.

27 Binder G, Keller E, Mix M, Massa GG, Stokvis-Brantsma WH, Wit JM, Ranke MB: Isolated GH deficiency with dominant inheritance: new mutations, new insights. J Clin Endocrinol Metab 2001;86:3877–3881.

28 Hayashi Y, Yamamoto M, Ohmori S, Kamijo T, Ogawa M, Seo H: Inhibition of growth hormone (GH) secretion by a mutant GH-I gene product in neuroendocrine cells containing secretory granules: an implication for isolated GH deficiency inherited in an autosomal dominant manner. J Clin Endocrinol Metab 1999;84:2134–2139.

29 Salemi S, Yousefi S, Lochmatter D, Eble A, Deladoey J, Robinson IC, Simon HU, Mullis PE: Isolated autosomal dominant growth hormone deficiency: stimulating mutant GH-1 gene expression drives GH-1 splice-site selection, cell proliferation, and apoptosis. Endocrinology 2007;148:45–53.

30 Lochmatter D, Mullis PE: RNA interference in mammalian cell systems. Horm Res Paediatr 2011;75:63–69.

31 Garcia-Blanco MA, Baraniak AP, Lasda EL: Alternative splicing in disease and therapy. Nat Biotechnol 2004;22:535–546.

32 Shariat N, Ryther RC, Phillips JA 3rd, Robinson IC, Patton JG: Rescue of pituitary function in a mouse model of isolated growth hormone deficiency type 2 by RNA interference. Endocrinology 2008;149:580–586.

33 Fedorov Y, Anderson EM, Birmingham A, Reynolds A, Karpilow J, Robinson K, Leake D, Marshall WS, Khvorova A: Off-target effects by siRNA can induce toxic phenotype. RNA 2006;12:1188–1196.

34 Grimm D, Streetz KL, Jopling CL, Storm TA, Pandey K, Davis CR, Marion P, Salazar F, Kay MA: Fatality in mice due to oversaturation of cellular microRNA/short hairpin RNA pathways. Nature 2006;441:537–541.

35 Boudreau RL, Martins I, Davidson BL: Artificial microRNAs as siRNA shuttles: improved safety as compared to shRNAs in vitro and in vivo. Mol Ther 2009;17:169–175.

36 Castanotto D, Sakurai K, Lingeman R, Li H, Shively L, Aagaard L, Soifer H, Gatignol A, Riggs A, Rossi JJ: Combinatorial delivery of small interfering RNAs reduces RNAi efficacy by selective incorporation into RISC. Nucleic Acids Res 2007;35:5154–5164.

37 Naldini L: Lentiviruses as gene transfer agents for delivery to non-dividing cells. Curr Opin Biotechnol 1998;9:457–463.

38 Marquis J, Kampfer SS, Angehrn L, Schumperli D: Doxycycline-controlled splicing modulation by regulated antisense U7 snRNA expression cassettes. Gene Ther 2009;16:70–77.

39 Brown BD, Cantore A, Annoni A, Sergi LS, Lombardo A, Della Valle P, D'Angelo A, Naldini L: A microRNA-regulated lentiviral vector mediates stable correction of hemophilia B mice. Blood 2007;110:4144–4152.

40 Cavazzana-Calvo M, Hacein-Bey S, de Saint Basile G, Gross F, Yvon E, Nusbaum P, Selz F, Hue C, Certain S, Casanova JL, Bousso P, Deist FL, Fischer A: Gene therapy of human severe combined immunodeficiency (SCID)-X1 disease. Science 2000;288:669–672.

41 Stegmeier F, Hu G, Rickles RJ, Hannon GJ, Elledge SJ: A lentiviral microRNA-based system for single-copy polymerase 2-regulated RNA interference in mammalian cells. Proc Natl Acad Sci USA 2005;102: 13212–13217.

42 Lochmatter D, Strom M, Eble A, Petkovic V, Fluck CE, Bidlingmaier M, Robinson IC, Mullis PE: Isolated GH deficiency type 2: knockdown of the harmful Delta3GH recovers wt-GH secretion in rat tumor pituitary cells. Endocrinology 2010;151: 4400–4409.

43 Nissim-Rafinia M, Kerem B: The splicing machinery is a genetic modifier of disease severity. Trends Genet 2005;21:480–483.

44 Liu HX, Zhang M, Krainer AR: Identification of functional exonic splicing enhancer motifs recognized by individual SR proteins. Genes Dev 1998;12: 1998–2012.

45 Birney E, Kumar S, Krainer AR: Analysis of the RNA-recognition motif and RS and RGG domains: conservation in metazoan pre-mRNA splicing factors. Nucleic Acids Res 1993;21:5803–5816.

46 Zhu J, Mayeda A, Krainer AR: Exon identity established through differential antagonism between exonic splicing silencer-bound hnRNP A1 and enhancer-bound SR proteins. Mol Cell 2001;8: 1351–1361.

47 Hastings ML, Krainer AR: Pre-mRNA splicing in the new millennium. Curr Opin Cell Biol 2001;13: 302–309.

48 Cartegni L, Chew SL, Krainer AR: Listening to silence and understanding nonsense: exonic mutations that affect splicing. Nat Rev Genet 2002;3: 285–298.

49 Ma X, Ezzeldin HH, Diasio RB: Histone deacetylase inhibitors: current status and overview of recent clinical trials. Drugs 2009;69:1911–1934.

50 Nissim-Rafinia M, Kerem B: Splicing regulation as a potential genetic modifier. Trends Genet 2002;18: 123–127.

51 Chang JG, Hsieh-Li HM, Jong YJ, Wang NM, Tsai CH, Li H: Treatment of spinal muscular atrophy by sodium butyrate. Proc Natl Acad Sci USA 2001;98: 9808–9813.

52 Nissim-Rafinia M, Aviram M, Randell SH, Shushi L, Ozeri E, Chiba-Falek O, Eidelman O, Pollard HB, Yankaskas JR, Kerem B: Restoration of the cystic fibrosis transmembrane conductance regulator function by splicing modulation. EMBO Rep 2004; 5:1071–1077.

53 Niel F, Legendre M, Bienvenu T, Bieth E, Lalau G, Sermet I, Bondeux D, Boukari R, Derelle J, Levy P, Ruszniewski P, Martin J, Costa C, Goossens M, Girodon E: A new large CFTR rearrangement illustrates the importance of searching for complex alleles. Hum Mutat 2006;27:716–717.

54 Heerdt BG, Houston MA, Anthony GM, Augenlicht LH: Initiation of growth arrest and apoptosis of MCF-7 mammary carcinoma cells by tributyrin, a triglyceride analogue of the short-chain fatty acid butyrate, is associated with mitochondrial activity. Cancer Res 1999;59:1584–1591.

Maria Consolata Miletta
University Children's Hospital, Pediatric Endocrinology, Inselspital
CH–3010 Bern (Switzerland)
Tel. +41 31 632 9557, E-Mail maria.miletta@dkf.unibe.ch

Author Index

Subject Index